TABLE OF CONTENTS

Pocket Guide to accompany

Chemistry
&Chemical Reactivity

THIRD EDITION

Kotz & Treichel

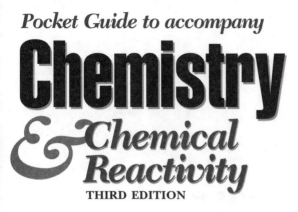

John M. DeKorte

Glendale Community College

Saunders Golden Sunburst Series
Saunders College Publishing
Harcourt Brace College Publishers

Fort Worth Philadelphia San Diego New York Orlando Austin
San Antonio Toronto Montreal London Sydney Tokyo

Printed in the United States of America.

DeKorte; Poctet Guide to accompany Chemistry & Chemical
Reactivity, 3E. Kotz and Treichel.

ISBN 0-03-017562-3

567 095 987654321

PREFACE FOR STUDENTS

This study guide is intended to be a synopsis of the third edition of the highly successful general chemistry textbook "Chemistry and Chemical Reactivity" by Kotz and Treichel. The guide has condensed coverage of each section of the main text, followed by Sections entitled "Key Expressions" and "Hints for Answering Questions and Solving Problem" for each chapter. Every attempt has been made to present the material with a clear, crisp writing style that will enable you to grasp quickly the essential portions of each topic. General guidelines for solving problems are given throughout the guide, but no attempt has been made to incorporate examples that would duplicate those found in the text.

The pedagogical features of the guide include the use of:

- **boldface type** to highlight terms in sentences in which they are being defined,

- *boldface italics to highlight key concepts and sentences*, sentences that should normally be highlighted using a colored highliter, *though only introductory sentences associated with sections of important indented material are highlighted and highlighting is not used with equations,*

- ✳ with change of font to point out things you can try to improve your performance in chemistry,

- "bulleted" formats like this to enable you to identify quickly important concepts and steps to follow in problem solving, and

- labels with important expressions for quick reference. These labels have the format of (Chapter number.letter), as in (6b), followed by (Chapter number.number) when the expression also corresponds to a numbered equation in the main text, as in (6c,6.2).

All other references to material in the main textbook clearly state "in the text."

The guide is published in a convenient size to enable you to take it with you wherever you go and thus take advantage of opportunities to study that might otherwise be lost. Consistent use of the guide should increase the rate at which you can assimilate terms and concepts and greatly improve your under-standing of chemistry. However, chemistry is also a quantita-tive science, and there is no substitute for practicing problem solving. Hence, you will need to spend adequate time practicing solving problems to be successful in general chemistry. Using the clear steps and hints of the guide to identify and solve problems should enhance your problem solving ability and overall chance for success.

We think you will find this guide to be very helpful in your study of general chemistry and wish you the best as you discover the power of chemistry to interpret the world in which we live. We have tried to keep the guide as error free as possible but probably have not succeeded completely. Any errors remaining are the sole responsibility of the author. Please let us know about any errors or problems, as well as ideas that you might have for improving the guide in future editions.

The author appreciates the cooperation of the lead author of the main textbook, John Kotz, and John Vondeling and Beth Rosato of Saunders College Publishing. A special word of thanks goes to my wife Carolyn for her support and understanding during the preparation of this study guide.

John M. DeKorte
Chemistry Department
Glendale Community College
Glendale, AZ 85302

INTRODUCTION

THE NATURE OF CHEMISTRY

Science And Its Methods

The systematic process that is often used in pursuing scientific investigations is known as the **scientific method**. An investigator poses a reasonable question or topic for investigation and searches the literature to learn what is known about the topic from the work of others. After forming a tentative explanation for the phenomena or putting forth a prediction of experimental results that might be found, a **hypothesis**, the investigator conducts experiments that are designed to give results that will either support or refute the hypothesis. In chemistry, this typically involves collecting both **qualitative** (nonnumerical) and **quantitative** (numerical) **information**.

The preliminary data can cause the investigator to modify the original hypothesis and conduct additional experiments. After determining that the results are consistently reproducible, the investigator attempts to identify patterns in the results and formulate a summary of the results in terms of a general rule. Further testing over extended periods of time can result in this rule becoming a **law**, a concise verbal or mathematical statement of a relation that seems always to be the same under the same conditions.

The next step in the process involves conceiving a **theory,** a unifying principle that explains a body of facts and the laws based on them. A theory can suggest new hypotheses and experiments. Keep in mind, however, that theories are inventions of the human mind and can and do change as new facts are uncovered.

In other cases, scientific discoveries are the results of lucky accidents occurring for people with prepared minds. Discoveries of this kind are said to be made serendipitously and include the discovery of penicillin, teflon and the cancer drug known as cisplatin. The discovery of cisplatin by Dr. Barnett Rosenberg is described in detail in the text.

Risks and Benefits

There are risks and benefits associated with most discoveries. "Risk assessment" is a process that brings together people from various sciences to establish the severity of the health risk associated with exposure to a particular compound. "Risk management", on the other hand, involves ethics, economics and other matters that are part of government and politics. In recent years, a new discipline called "risk communication" has emerged to promote the accurate transfer of information between the scientists who assess the risk, the government agencies that manage the risk and the public.

You will need to make many decisions concerning risks and benefits throughout your lifetime and therefore will have many opportunities to use your understanding of science in general and chemistry in particular. Interestingly, it has been determined that people usually are more willing to accept voluntary risks than involuntary risks. Hence, one of the goals of scientists is to create a scientifically literate society which can make informed and intelligent decisions.

CHAPTER 1

MATTER AND MEASUREMENT

This chapter describes some of the properties of matter and the types of changes it can undergo. The chapter then describes the system of measurements that is used in chemistry, the procedure that is used for recording measurements in a way that reflects the certainty of the measurement and the procedures for correctly converting between units when using measured quantities in calculations.

1.1   Physical Properties

Matter is anything that occupies space and has mass. **Physical properties** are properties which can be observed and measured without changing the composition of a substance. Table 1.1 in the text lists some physical properties of matter that chemists commonly use. ✳ Try preparing a "flash card" listing these physical properties and memorizing them as examples of physical properties..

States of Matter

Matter commonly exists in three physical states: the solid, liquid and gaseous states. **Solids** have rigid shapes and fixed volumes that change very little as temperature and pressure change. **Liquids** also have a fixed volume but have no fixed shape of their own and therefore take on the shapes of their containers. **Gases** expand to fill whatever container they occupy, and their volume varies considerably with temperature and pressure.

✳ Try preparing three separate "flash cards" summarizing the characteristics of these three states of matter and memorizing them.

Kinetic-Molecular Theory

The physical properties described in the text refer to samples of matter that are large enough to be seen, measured and handled. Such samples are called **macroscopic** samples, in contrast to **microscopic** samples, such as cells and microorganisms. The structure of matter that is of most interest to chemists, however, is the **submicroscopic** scale of atoms and molecules. One theory that helps us interpret the physical properties of matter at this level is the **kinetic-molecular theory (KMT) of matter**.

According to the KMT, all matter is composed of atoms or molecules, which are in constant motion. The higher the temperature, the faster the particles move. Thus, the higher the temperature, the higher the kinetic energy of the particles. The increase in kinetic energy with increasing temperature contributes to changing substances from the solid to liquid to gaseous states, though *we shall see later that all changes of state occur at constant temperatures*. Heat transferred to effect these changes of state enables particles to overcome the attractions of neighboring particles and move away from each other. This involves changing the positions of the atoms and molecules and therefore changing their **potential energy**.

Density and Temperature

Density, given by the ratio of the mass of an object to its volume, is a physical property that is useful for identifying objects.

$$\text{Density} = \frac{\text{mass}}{\text{volume}} \qquad \text{(1a)}$$

The density of an object can be determined by measuring the mass of the object using a balance and by measuring the volume of the object by the increase in volume that is caused by carefully adding the object to water that is in a graduated cylinder.

If any two of the three quantities-mass, volume and density-are known for a sample of matter, they can be used to calculate the third. Ex. 1.1 in the text illustrates the use of volume and density to calculate mass.

The temperature at which the solid form of a pure element or compound melts (its **melting point**) or the liquid form boils (its **boiling point**) are additional physical properties that are useful for identifying objects. **Temperature** is the property of matter that determines whether heat energy can be transferred from one body to another and the direction of the heat flow: *heat energy transfers spontaneously from hotter objects to cooler objects*.

The actual number that is used to represent the temperature of an object depends on the unit chosen for the measurement. Three temperature scales commonly used are: the Fahrenheit, Celsius and Kelvin scales. *A Celsius degree is 1.8 times as large as a Fahrenheit degree and the same size as a kelvin; the word "degree" is not used with the SI unit of kelvins*. Zero on the Kelvin scale is the lowest possible temperature, a point called **absolute zero**. The equations for converting between temperature scales are:

$$t(°C) = \frac{°F - 32°}{1.8} \tag{1b}$$

and

$$T(K) = t(°C) + 273.15 \tag{1c}$$

When temperature data are used in calculations in the text, the temperature data are usually expressed in kelvins.

Physical and Chemical Changes

Changes involving physical properties are called **physical changes**. The chemical composition of the substance is preserved even though it may change its size, shape or physical state. For example, the melting of magnesium metal does not change the chemical composition of the magnesium; we expect to obtain solid magnesium metal on cooling. On the other hand, the burning of magnesium produces magnesium oxide, as well a bright white light that was the source of light obtained from "flash cubes" with cameras. There is a change in chemical composition, and the change is called a **chemical change**.

1.2   Elements and Atoms

Elements are pure substances that are composed of only one type of atom. About 90 of the 111 elements known at this time are found in nature; the remainder were created by scientists in nuclear reactions. *Each element has a name and a symbol, which is an abbreviation of its English or Latin name; the only exception being tungsten, which is represented by W from its German name of wolfram. If the symbol is composed of two letters, only the first letter is capitalized.*

An **atom** is the smallest particle of an element that retains the properties of that element. Modern biology and chemistry are based on exploring nature at the atomic and molecular level.

1.3   Compounds and Molecules

Chemical compounds are pure substances that are composed of two or more elements in fixed and definite proportions by mass. *When elements become part of a compound, their original properties are replaced by those of the compound.*

Some compounds are composed of **ions**, electrically charged atoms or groups of atoms. Other compounds exist as **molecules**, which are the smallest discrete units that retain the properties of the compound. The composition of any compound can be represented by its **formula**. The formula NaCl tells us that sodium chloride is composed of Na^+ and Cl^- ions in a 1:1 ratio whereas the formula $C_{12}H_{22}O_{11}$ tells us that each molecule of sucrose is composed of 12 C atoms, 22 H atoms and 11 O atoms. Many physical and chemical properties depend on **molecular structures**, which refer to three-dimensional arrangements of atoms in molecules. You will learn how to identify ionic and molecular covalent compounds and will be exploring relations between structures and properties in future chapters.

1.4   Chemical and Physical Changes

In **chemical changes** or **chemical reactions**, one or more of the reacting substances, known as **reactants,** are transformed into one or more different substances, known as products. *The reactants are written on the left side of the chemical equation and the products on the right side; they are separated by an arrow that points to the right and can be interpreted to mean "produces," "yields" or "forms." At the molecular level chemical changes produce a new arrangement of atoms without any loss or gain of atoms; in other words, atoms are merely rearranged during chemical reactions.*

Chemical reactions are represented by **chemical equations**. The chemical equation $H_2(g) + Cl_2(g) \rightarrow 2\ HCl(g)$ represents the reaction of one molecule of gaseous hydrogen with one molecule of gaseous chlorine to form 2 molecules of gaseous hydrogen chloride. The hydrogen and chlorine are the reactants, and the hydrogen chloride is the product. There are two atoms of hydrogen and chlorine on each side of the equation but the arrangement of these atoms in the products differs from that in the reactants. You will learn more about writing chemical equations when studying Chapter 4 of the text.

The **chemical properties** of a substance are those properties that involve chemical changes. ✶ Try preparing a "flash card" listing the chemical properties that are given in the text and memorizing them as examples of kinds of chemical changes.

1.5   Mixtures and Pure Substances

A **heterogeneous mixture** is composed of two or more substances and often has variable composition throughout. There often is a distinct boundary between the components of a heterogeneous mixture, as with iron filings and sugar and with an ice cube and liquid water. A **homogeneous mixture**, or **solution**, consists of two or more substances in the same phase and has the same composition throughout.

The properties of a heterogeneous mixture can vary from one region of a given sample to another whereas the properties of a homogeneous mixture are the same throughout. The properties of homogeneous mixtures can, however, vary from one sample to another depending on the amount of each component present. Thus, the properties of a dilute solution of sugar in water can vary from those of a concentrated solution of sugar in water.

A mixture can be separated into its components by physical means. When a mixture is separated into its pure components, the components are said to be *purified.* We define a **pure substance** as a sample of matter with properties that cannot be changed by further purification.

1.6 Units of Measurement

The scientific community has chosen a modified version of the **metric system** as the standard system for recording and reporting measurements, the **SI** system. This is a decimal-based system in which all of the units are expressed as powers of 10 times some basic unit.

SI Units

The six most commonly used SI base units are listed in Table 1.3 in the text, and the eight most commonly used prefixes are listed in Table 1.4 in the text. ✶ Try preparing two separate "flash cards" containing this information and memorizing them. Also become familiar with the conversion factors given in Table 1.5, because you will need to refer this table often.

The SI unit of volume is a derived unit, the cubic meter (m^3), and is too large for everyday laboratory use. We use the unit called the **liter**, symbolized by **L**: $1L = 1,000$ mL $= 1,000$ cm^3 $(1,000$ cc$)$.

1.7 Using Numerical Information

The text emphasizes the use of **dimensional analysis** for converting from one set of units to another. *The quantity that is to be converted is multiplied by a conversion factor to produce the same quantity expressed in new units.*

Quantity expressed · conversion factor = Quantity expressed
in current units   in new units   (1d)

Number, current unit · <u>   number, new unit   </u> = new number, new unit
               number, current unit

Conversion factors are always written in the form that causes current units to cancel and gives new units. If the current units are in the numerator, this requires using (new units divided by current units). If the current units are in the denominator, this requires using (current units divided by new units). Exs. 1.2-5 in the text illustrate the proper use of conversion factors.

Significant Figures

The **precision** of a measurement indicates how well several determinations of the same quantity agree with themselves. **Accuracy** indicates how well measured and determined quantities agree with accepted or actual values.

When we report a measured quantity we usually report all the digits that are known with certainty plus the first digit in which there is uncertainty; an uncertainty of ±1 is normally assumed for the last digit reported, unless stated otherwise. These digits are known collectively as **significant figures**.

Guidelines for Determining and Using Significant Figures

The text gives four rules for determining the numbers of significant figures in measured quantities and using measured quantities in calculations. *Notice that:*

• Zeros which are used to show the decimal place are never counted as significant figures. This means that zeros that are to the left of the first nonzero digit are never significant; 0.0012 m has only two "sig figs."

• Zeros that are ending zeros and to the left of the decimal point will only be considered to be significant when followed by a decimal point; 50 has one sig fig whereas 50. has two sig figs.

• Zeros that are ending zeros and to the right of the decimal point are always counted as sig figs; 0.0520 g has three sig figs, the five, two and ending zero.

• Calculated quantities simply cannot be any more precise than the least precise piece of information that goes into the calculations.

- In terms of adding and subtracting, this means no more decimal places in the answer than there are in the quantity with the least number of decimal places.

- In terms of multiplying and dividing, this means no more total sig figs in the answer than there total sig figs in the quantity having the fewest number of <u>total</u> sig figs.

- Counted and defined quantities are considered to have an unlimited number of sig figs, and thus, do not affect the appropriate number of sig figs for answers.

- You should complete your calculations by using all the digits allowed by your calculator and only round to the correct number of sig figs at the end of the calculation.

- If the first digit to be discarded is a 5 followed by other digits or is greater than 5, round up; round both 0.6753 and 0.676 to 0.68 to two sig figs.

- If the first digit to be discarded is a 5 and is not followed by other digits, round the preceding digit to an even number; round 0.935 to 0.94 to two sig figs and 0.885 to 0.88 to two sig figs.

- If the first digit to be discarded is less than 5, round down; round 0.0.834 to 0.83 to two sig figs.

✶ Try preparing separate "flash cards" for each rule and memorizing them. Also try working with them until you do not have to stop too long to think about them.

The Concept of Percent

Percent is a shortened version of the Latin phrase *per centum*, meaning in 100. Thus, ***percent can always be used to write a conversion factor in which the value of the percent is divided by 100.*** For example, a 30.0% by mass aqueous solution of hydrogen peroxide, H_2O_2, contains 30.0 g H_2O_2/100.0 g solution. This also means there are 100.0 g solution/30.0 g H_2O_2, 30.0 g H_2O_2/70.0 g H_2O and 70.0 g H_2O/30.0 g H_2O. Ex. 1.7 in the text also illustrates the use of percent to write conversion factors.

1.8   Key Expressions

Density expression:

$$d = \frac{mass}{volume} = \frac{m}{v}$$

Temperature conversion expressions:

$$t(^{\circ}C) = \frac{^{\circ}F - 32^{\circ}}{1.8}$$

and

$$T(K) = t(^{\circ}C) + 273.15$$

1.9   Hints for Answering Questions and Solving Problems

Writing Conversions Factors:

It is important to realize that all relations can be used to write two conversion factors. For example, $1 \text{ mm} = 10^{-3} \text{ m}$ can be used to write

$$\frac{1 \text{ mm}}{10^{-3} \text{ m}} \qquad \text{and} \qquad \frac{10^{-3} \text{ m}}{1 \text{ mm}}$$

because we can divide either side of the $1 \text{ mm} = 10^{-3} \text{ m}$ equality by the other to obtain these ratios and show that they are equal to one. Similarly, a density of 13.6 g/cm^3 for Hg can be used to write

$$\frac{13.6 \text{ g Hg}}{1 \text{ cm}^3} \qquad \text{and} \qquad \frac{1 \text{ cm}^3}{13.6 \text{ g Hg}}$$

and both conversion factors should be considered to have 3 sig figs. Lastly, refer to 1.7 to recall how percent can be used to write conversion factors.

Setting Up Unit Conversion Problems:

Some students find it helpful to begin each unit conversion problem by writing a ? and a statement of what is being asked, followed by an equals sign and the key piece of information that is given; that is, by writing what is wanted and what is given. Consider the following examples:

• Example 1. The length of an olympic size swimming pool is 50.0 meters. Calculate the length of an olympic size swimming pool in feet.

The statement of the problem asks us to calculate the length of the pool in feet given its length in meters, so we could start by writing

$$? \text{ length in ft} = 50.0 \text{ m} \cdot$$

This would be followed by looking at Table 1.5 to obtain the necessary conversion factors for converting m to ft.

The path that is suggested by the relations available in Table 1.5 is:

$$m \rightarrow cm \rightarrow in \rightarrow ft$$

Thus, we could write

$$? \text{ length in ft} = 50.0 \text{ m} \cdot \frac{100 \text{ cm}}{1 \text{ m}} \cdot \frac{1 \text{ in}}{2.54 \text{ cm}} \cdot \frac{1 \text{ ft}}{12 \text{ in}}$$

and express our answer to three sig figs as 164 ft. This is reasonable, since 1 m is ≈10% longer than 1 yd, is ≈3.3 ft.

• Example 2. If 5.84×10^{13} (spherical) neon atoms were to be laid in a straight line, each touching the next, the line would measure 5.15 miles. Calculate the width (diameter) of a neon atom in nanometers.

The statement of the problem asks us to calculate the width of a neon atom in nanometers, the nanometers/atom, so we could start by writing

$$? \frac{nm}{atom} =$$

and therefore note that we will need to write the given information as

$$? \frac{nm}{atom} = \frac{5.15 \text{ mi}}{5.84 \times 10^{13} \text{ atoms}} \cdot$$

The path that is suggested for converting mi to nm by the relations in Table 1.5 is:

$$mi \rightarrow km \rightarrow m \rightarrow nm$$

Thus, we could write

$$? \frac{nm}{atom} = \frac{5.15 \text{ mi}}{5.84 \times 10^{13} \text{ atoms}} \cdot \frac{1 \text{ km}}{0.62137 \text{ mi}} \cdot \frac{10^3 \text{ m}}{1 \text{ km}} \cdot \frac{1 \text{ nm}}{10^{-9} \text{ m}}$$

and express our answer to three sig figs as 0.142 nm/atom.

> Note: Whenever you are asked to convert between two non-base units in the metric system, such as km to nm, it is wise to convert the given unit to the base unit before proceeding to convert to the desired unit. The conversion factors for going to the base unit and on to the desired unit are readily available whereas the conversion factor for the direct conversion is not, and it is easy to make a mistake when trying to construct the direct conversion factor.

• Example 3. A sprinter runs the 100 m dash in 10.3 s. Calculate the speed of the sprinter in mi/hr.

The statement of the problem asks us to calculate the speed of the sprinter in mi/hr, so we could start by building on what was learned in working the previous example, Example 2, and writing

$$? \; \frac{mi}{hr} = \frac{100 \; m}{10.3 \; s} \cdot$$

The path that is suggested for converting m to mi by the relations in Table 1.5 is:

$$m \rightarrow km \rightarrow mi$$

whereas the path for converting s to hr is:

$$s \rightarrow min \rightarrow hr$$

Thus, we could write

$$? \; \frac{mi}{hr} = \frac{100 \; m}{10.3 \; s} \cdot \frac{1 \; km}{10^3 \; m} \cdot \frac{0.62137 \; mi}{1 \; km} \cdot \frac{60 \; s}{1 \; min} \cdot \frac{60 \; min}{1 \; hr}$$

and express our answer to three sig figs as 21.7 mi/hr.

> Note: If the current units are in the numerator,
> this requires using (new units/current units) as a
> conversion factor. If, however, the current units
> are in the denominator, this requires using (current
> units/new units) as a conversion factor. This is why
> the unit m appears in the denominator and the unit
> s appears in the numerator of the chain of conver-
> sion factors used in working this example.

Conducting Unit Conversions:

There will be some instances in which it will be necessary to raise a conversion factor to a power to accomplish a desired unit conversion, as was done in Ex. 1.5 in the text. When this is the case, you should realize that the power merely tells us the number of times the conversion factor must be used. Thus, converting cm^3 to m^3 using

$$1 \text{ cm}^3 \cdot (\text{ 1 m/100 cm})^3$$

is equivalent to using

$$1 \text{ cm}^3 \cdot \frac{1 \text{ m}}{100 \text{ cm}} \cdot \frac{1 \text{ m}}{100 \text{ cm}} \cdot \frac{1 \text{ m}}{100 \text{ cm}}$$

with your calculator.

CHAPTER 2

ATOMS AND ELEMENTS

This chapter describes the origins of atomic theory, the discovery and characterization of subatomic particles and the locations of these particles in atoms, as well as the method for designating the composition of individual atoms. The existence of isotopes of elements and the impact of isotopes on atomic masses of elements are discussed. The arrangement of the elements in the periodic table is then described followed by the introduction of the mole as the quantity of matter that connects the submicroscopic world of atoms with the macroscopic world in which we live.

2.1   Origins Of Atomic Theory

The most important contributions to our current model of the atomic nature of matter were those of John Dalton (1766-1844). *In 1803, Dalton put forth the following postulates:*

• All matter is made of atoms. These indivisible and indestructible objects are the ultimate chemical particles.

• All atoms of a given element are identical, both in mass and in properties. Atoms of different elements have different masses and different properties.

• Compounds are formed by combination of two or more different kinds of atoms. Atoms combine in the ratio of small whole numbers, for example, one atom of A with one atom B, or two atoms of A with one atom of B.

• Atoms are the units of chemical change. A chemical reaction involves only combination, separation, or rear- rangement of atoms, but atoms are not created, destroyed, divided into parts, or converted into other kinds of atoms during a chemical reaction.

Dalton's ideas were readily accepted by the scientific community because they could be used to explain several scientific laws that were already known, the law of conservation of matter and the law of constant composition or definite proportions. In addition, they lead Dalton to postulate a new law, the law of multiple proportions. Consider the following summary of these

• The law of **conservation of matter**, based on the work of Antoine Lavoisier and others, states there is no observable change in mass during chemical reactions. This can explained by realizing there is a conservation of atoms of each kind during chemical reactions. See Postulates 2 and 4 of Dalton's Atomic Theory.

• The law of **constant composition** or **definite proportions**, based on the work of Joseph Proust, states that a particular compound, once purified, always contains the same elements in the same ratio by mass. This can be explained by realizing atoms combine in the ratio of small whole numbers. See Postulates 2 and 3 of Dalton's Atomic Theory.

• The law of **multiple proportions**, based on the work of John Dalton himself, states that when two elements form two different compounds, the mass ratio in one compound is a small whole number times the mass ratio in the other. This can be explained by realizing that atoms can combine in differing small whole number ratios, as in CO and CO_2 where the oxygen:carbon mass ratio in CO_2 is twice that in CO (2:1 mass ratio) and in SO_2 and in SO_3 where the oxygen: sulfur mass ratio in SO_3 is 3/2 times that in SO_2 (3:2 mass ratio). See Postulate 3 of Dalton's Atomic Theory.

✳ Try preparing three separate "flash cards" summarizing these laws and memorizing them. Also become familiar with the use of Dalton's postulates to rationalize these laws.

2.2 Atomic Structure

Electricity

Electricity was involved in many of the experiments from which the theory of atomic structure was derived. Benjamin Franklin was the one who named the two types of electrical charge as positive (+) and negative (−), because they were opposites and could neutralize each other. *Experiments with an electroscope show that like charges repel each other and opposite charges attract each other.*

Franklin also concluded that electrical charge was conserved, meaning that if a negative charge appeared somewhere, a positive charge of the same size must appear somewhere else. This lead scientists to believe that the buildup of charge that occurs when one substance is rubbed against another must cause a separation of charges and that these charges were somehow associated with matter-perhaps with atoms.

Radioactivity

In 1896, the French physicist Henri Becquerel discovered radioactivity in a fortuitous and serendipitous manner. Becquerel was studying the effects of sunlight on uranium ore and was using photographic plates to try to detect any rays that might be emitted. On a cloudy day, Becquerel wrapped a sample of uranium ore and a photographic plate together in black paper and placed the package in a drawer. He developed the stored plate without conducting the planned exposure to light and was surprised to obtain an image of the uranium ore sample. *The sample had spontaneously emitted rays to cause this to happen.* The phenomena was further studied by Madame Curie who discovered polonium and radium behaved in the same manner, and in 1899, named the phenomenon **radioactivity**.

Radioactive elements can spontaneously emit three kinds of radiation:

- alpha(α) particles which have an electrical charge of +2,

- beta(β) particles which have an electrical charge of -1 and are deflected by electrically charged plates more than alpha particles and must therefore be much lighter and

- gamma(γ) rays which have no detectable charge or mass and behave like light rays.

Madame Curie's suggestion that some atoms can spontaneously disintegrate by emitting radiation contradicted Dalton's idea that atoms were indivisible and therefore required an extension of Dalton's theory be made. If atoms can break apart, they must be composed of something smaller than atoms, they must be composed of subatomic particles.

Electrons

Passing an electrical current through a solution of a compound can cause a chemical reaction to occur in a process called **electrolysis;** electrolysis is used for producing aluminum metal, purifying copper metal to use in electrical wiring and chrome plating automobile parts. In 1833, Michael Faraday studied the quantitative aspects of electrolysis and came to the conclusion that there was a fundamental particle of electricity which was given the name **electron**.

Further evidence for the existence of electrons was obtained by using cathode ray tubes, which were the forerunners of tubes used in television sets and computer monitors. *In 1897, J.J. Thomson used electrical and magnetic fields to cause offsetting deflections of cathode rays to determine the ratio of charge-to-mass (e/m) for the cathode ray particles. Thomson obtained the same e/m ratio using different metal cathodes and different gases in the tubes and came to the conclusion that cathode ray particles were a fundamental particle of matter and identical to the particles suggested by the work of Faraday;* the negative electrical charge of the cathode ray particles having been established by the direction of travel (negative to positive electrode) and the direction of deflection caused by the external electrical and magnetic fields.

The charge on the electron was determined later by the American physicist Robert Millikan using an oil-drop experiment. X rays were used to dislodge electrons from gaseous atoms, and the electrons became attached to oil drops which naturally began to fall due to gravity. Millikan adjusted the electric charge on plates above and below the drops and was able to suspend individual drops and thus determine the amount of charge on individual drops. *He found that the charge was always a multiple of -1.6 x 10^{-19} C, where C represents the SI unit of electrical charge, the coulomb.* This value was then used with the e/m ratio obtained by Thomson to obtain the value of the mass of the electron. *The currently accepted values for charge and mass for an electron are -1.60217733 x 10^{-19} C and 9.109389 x 10^{-28} g, though we refer to the charge as being just -1 as a matter of convenience.*

Protons

The first experimental evidence for a fundamental positive particle came from cathode ray tube studies that used cathodes with holes in them; see Fig. 2.7 in the text. The positively charged particles were formed as cathode ray particles collided with the gas molecules in the tube causing them to lose electrons. The source of the positive particles was confirmed by the observation that the charge-to-mass ratio (e/m) of the positive particles changed with the kind of gas that was being used in the tube. *The largest e/m ratio was obtained using hydrogen gas, and other e/m ratios were always multiples of this value suggesting H$^+$ is a fundamental positive particle.* These particles were called **protons** (from a Greek word meaning "the primary one") by Ernest Rutherford. *The currently accepted values for charge and mass for a proton are 1.60217733 x 10^{-19} C and 1.672623 x 10^{-24} g, and we refer to the charge as being just +1 as a matter of convenience.*

Neutrons

Atoms are neutral and therefore must contain equal numbers of protons and electrons, but their masses are generally greater than would be predicted on the basis of only protons and electrons, so they also must contain uncharged particles called **neutrons**. In 1932, neutrons were discovered by James Chadwick using nuclear reaction techniques. The neutron has no charge and a mass of $1.6749286 \times 10^{-24}$ g, which is nearly identical to that of a proton and approximately 1840 times greater than the mass of an electron.

The Nucleus of the Atom

J.J. Thomson postulated the atom was a uniform sphere of positively charged matter in which electrons circulate in coplanar rings. Thomson's students tested this model by directing a beam of electrons at a thin metal foil and observed a smaller than expected deflection. Thus, Thomson lowered his estimate of the numbers of electrons in atoms, but did not revise his model of the atom.

Ernest Rutherford later aimed alpha particles, which were then known to be positively charged particles having the mass of helium atoms, at a thin sheet of gold foil to test the model of Thomson, his former mentor. Rutherford reasoned that such massive particles would be deflected very little, if Thomson's model were correct. However, Rutherford, his associate Hans Geiger and a young student, Ernest Marsden, were very much surprised by the fact that some of the α particles were deflected through large angles and a few came almost straight back. Thus, *they were forced to conclude that all of the positive charge and most of the mass of the atom is concentrated in a very small volume*, which Rutherford called the **nucleus**. The electrons occupy the rest of the space in the atom, and atoms are approximately 100,000 times larger than nuclei.

2.3   Atomic Composition

The three primary components of atoms are electrons, protons and neutrons. The protons and neutrons are located in the nucleus and account for all of the positive charge and almost all of the mass of an atom. The electrons occupy most of the volume of an atom but contribute very little mass. Because atoms are electrically neutral, the number of electrons outside the nucleus must equal the number of protons inside the nucleus.

All atoms of the same element have the same number of protons in the nucleus. This number, called the **atomic number** of the element, is given the symbol **Z** and is shown above the symbol of the element in the periodic table.

The masses of atoms are expressed relative to the mass of a carbon atom that has 6 protons and 6 neutrons in its nucleus. An atom of this type is defined to have a mass of exactly 12 **atomic mass units** (12 amu), so **1 amu** = 1/12th the mass of a carbon atom having 6 protons and 6 neutrons in its nucleus, 1/12th the mass of a so-called carbon-12 atom. The masses of all other atoms are expressed relative to the mass of this atom.

The masses of the proton and neutron are approximately 1 amu whereas electrons are about 1840 times lighter. Thus, we only need to add up the number of protons and neutrons in an atom to estimate its mass. The result is called the **mass number** of the particular atom and is given the symbol **A**, though you should always keep in mind that this number corresponds to the total number of protons and neutrons and not the actual mass of the atom. ✶ The characteristics of the subatomic particles of primary interest to chemists are summarized in Table 2.1. Try preparing a "flash card" summarizing the masses in amu, the charges and the symbols for these particles and memorizing them, calling the mass of the electron approximately 0.00055 amu and the masses of the proton and neutron approximately 1 amu. Also note the resemblance of the symbols used for subatomic particles to those described for atoms in the following paragraph.

We use the symbolism $^{A}_{Z}X$ to indicate the composition of an atom, where X is the symbol of the element, Z is its atomic number (its number of protons) and A is its mass number (its number of protons plus neutrons). Thus, the carbon atom that is the standard for all masses is represented by $^{12}_{6}C$ or simply by ^{12}C, because the symbol of the element tells us what the atomic number must be, and is called carbon-12. Similarly, the symbol $^{17}_{8}O$ is used to represent an atom of oxygen that is composed of 8 protons, 8 electrons and 9 (A - Z = 17 - 8 = 9) neutrons.

The masses of atoms relative to exactly 12 amu for carbon-12 can be determined by using mass spectrometers; see Fig. 2.11 in the text. *The masses of all atoms other than $^{1}_{1}H$ are slightly less than the estimate that is given by their mass numbers, because some mass is converted to energy binding the nucleus together.* This will be discussed further in Chapter 24 of the text. However, you should realize that atoms of $^{1}_{1}H$ only have one proton in the nucleus and therefore do not fall into this category. Interestingly then, *the most abundant atoms in the universe, $^{1}_{1}H$ atoms, are the only atoms which do not contain neutrons.*

2.4 Isotopes

Isotopes are atoms having the same atomic number Z but a different mass number A. In other words, *isotopes are atoms of the same element, but they have different masses because they have different numbers of neutrons.* Hydrogen, for example, exists as three isotopes: *protium*, ^1_1H; *deuterium* (sometimes represented by the symbol D), ^2_1H; and radioactive *tritium* (sometimes represented by the symbol T), ^3_1H. Most elements have at least two stable (nonradioactive) isotopes but some have only one. Aluminum, fluorine and phosphorus are examples of common elements having only one stable isotope.

The discovery of the existence of isotopes is another reason that Dalton's theory needed to be extended, because Dalton believed all the atoms of a given element had the same mass. See Postulate 2 of Dalton's Atomic Theory.

2.5 Atomic Mass

Isotope abundances and isotope masses can be determined using a mass spectrometer. Once the percent abundances and isotopic masses of each isotope in a naturally occurring sample of an element are known, they can be used to calculate the average mass of atoms of that element. The average mass, expressed in atomic mass units, is called the **atomic mass** of an element. *These are the masses that are shown below the symbols of the elements in the periodic table. In the case of radioactive elements, the mass number of the most stable isotope is given in parentheses in place of an atomic mass.*

To calculate the atomic mass of an element, we multiply the mass of each isotope by its percent abundance divided by 100, because percent means parts per hundred and gives the number of atoms of an isotope per hundred atoms total, *and sum the values obtained:*

Atomic mass = $\dfrac{\% \text{ abundance of isotope 1}}{100} \cdot$ mass of isotope 1 +

$\qquad\qquad \dfrac{\% \text{ abundance of isotope 2}}{100} \cdot$ mass of isotope 2 + ...

(2a)

This procedure is illustrated for bromine in Ex. 2.3 in the text. Also see Problem Solving Tips 2.1 in the text, and notice that the procedure for calculating atomic masses is totally analogous to the procedure for calculating your average score for your quizzes in a course or the average score for an examination for a class.

2.6   The Periodic Table

Features of the Periodic Table

Elements are listed in an order of increasing atomic number in the periodic table. They also are arranged in such a way that elements having similar chemical and physical properties lie in vertical columns called **groups**. The elements in groups designated as A groups in periodic tables used in the United States are classified as **main group elements** and those in B groups are classified as **transition elements**. The horizontal rows of the periodic table are called **periods**, and they are numbered beginning with 1 for the period containing only H and He. Thus, chlorine (Cl) is a main group element that can be found in Group VIIA and in the third period.

The elements that are to the left of the staircase line that starts between B and Al are classified as **metals** whereas those to the right of the line are classified as **nonmetals** and those along the line that are intermediate in character between metals and nonmetals are classified as **metalloids**. Metals are shown with blue shading in Fig. 2.13 in the text and in the periodic table in the front of the text, metalloids are shown in green and nonmetals are shown in yellow.

Metals are solids at room temperature and pressure (except for mercury), conduct electricity, are ductile (can be drawn into wires), are malleable (can be rolled into sheets) and can form alloys, which are solutions of one or more metals in another metal. Nonmetals, on the other hand, can exist as solids, liquids or gases at room temperature and pressure and are nonconductors of electricity (except for carbon as graphite). Metalloids exhibit characteristics of both metals and nonmetals; some are semiconductors of electricity that have been the basis of the electronics advances of the past several decades. ✳ Try preparing "flash cards" listing the general properties of metals and nonmetals and memorizing them.

Historical Development of the Periodic Table

The historical development of the periodic table illustrates the way chemistry has developed: experimental observations lead to empirical correlations of properties and then to the prediction of results of further experiments. The primary contributor to the modern periodic table was Dimitri Mendeleev (1834-1907), a professor of chemistry in Russia. Mendeleev arranged the known elements in an order of increasing atomic mass and in vertical groups of elements having similar chemical reactivity. He correctly placed Te and I in Groups VIA and VIIA on the basis of their chemical reactivity but was so convinced that chemical periodicity was a function of atomic mass, that he erroneously concluded their atomic masses (127.60 for Te and 126.90 for I) must be in error. However, we now know the atomic numbers of all the elements and realize that the elements should be arranged in an order of increasing atomic number, rather than an order of increasing atomic mass. Hence, Mendeleev's **law of chemical periodicity** has been revised to state that "the properties of the elements are periodic functions of their atomic numbers."

Chemistry and the Periodic Table

Several areas of the periodic table are given special names that are useful to know: Group IA, the **alkali metals**; Group IIA, the **alkaline earth metals**; the B group elements of the 4th-7th periods, the **transition elements**; Group VIIA, the **halogens;** Group VIII, the **noble gases;** the row of elements at the bottom following lanthanum, the **lanthanides** and the row of elements at the bottom following actinium, the **actinides**. ✶ Try preparing a "flash card" showing the shape of the periodic table and labelling these areas and memorizing them.

Chemical Periodicity

Elements in a group have similar, but not identical, properties. Indeed the properties of the first member of a group often differ from those of the lower members of the same group. However, members of a given group show a tendency to form similar compounds with other elements. For example, sodium and potassium from Group IA form NaCl and KCl, respectively, with chlorine, and chlorine and bromine form $CaCl_2$ and $CaBr_2$, respectively, with calcium and CCl_4 and CBr_4, respectively, with carbon. The reasons for both similarities and dissimilarities in chemical behavior will be discussed in later chapters.

2.7   The Mole: The Macro/Micro Connection

The chemical counting unit that has come into use to connect the submicroscopic world of atoms with the macroscopic world in which we live is the mole. *The mole, whose symbol is mol, is the SI unit for measuring an amount of substance.* A **mole** is the amount of substance that contains as many elementary entities (atoms, molecules or other particles) as there are atoms in exactly 12 g of carbon-12. *Experiments over the years have established that number as:*

$$1 \text{ mol} = 6.022136736 \times 10^{23} \text{ particles} \qquad (2b)$$

This value is commonly known as **Avogadro's number** in honor of Amedeo Avogadro (pronounced Av-o-god-ro) and is commonly used as 6.022×10^{23}. ✳ Try preparing a "flash card" to help you memorize this important number, because you will have many opportunities to use it throughout your study of chemistry.

Moles of Atoms, The Molar Mass

The mass in grams of 1 mol of atoms of any element ($6.022136736 \times 10^{23}$ atoms of that element) is the **molar mass** of that element. *Molar mass is abbreviated with a capital italicized M and is expressed in units of grams per mole (g/mol). It is numerically equal to the atomic mass expressed in atomic mass units. Thus,*

For sodium,

atomic mass:	22.9898 amu	(ave mass of 1 atom of Na)
molar mass:	22.9898 g	(mass of 6.022×10^{23} atoms of Na, mass of 1 mol of Na atoms)

For copper,

atomic mass:	63.546 amu	(ave mass of 1 atom of Cu)
molar mass:	63.546 g	(mass of 6.022×10^{23} atoms of Cu, mass of 1 mol of Cu atoms)

The mole concept is the cornerstone of quantitative chemistry. It is therefore imperative that you learn how to convert from moles to mass and from mass to moles using conversion factors.

To convert moles to mass, use

$$\text{moles} \cdot \frac{\text{grams}}{1 \text{ mol}} = \text{grams}$$

that is, moles · molar mass (2c)

To convert mass to moles, use

$$\text{grams} \cdot \frac{1 \text{ mol}}{\text{grams}} = \text{moles}$$

that is, grams · $\dfrac{1}{\text{molar mass}}$ (2d)

* Try preparing two separate "flash cards" summarizing these important unit conversions and memorizing them, because you will have many opportunities to use them throughout your study of chemistry. Also try using them until you are so familiar with them that you can use them without having to stop too long to think about them.

Exs. 2.5 and 2.6 in the text illustrate the use of the mole concept with elements. Notice the extension of these conversions to calculate numbers of atoms in Ex. 2.6; **the conversion between mass (grams) and moles always involves the use of molar mass and the conversion between particles and moles always involves the use of Avogadro's number.** See Problem Solving Tips and Ideas 2.2 in the text.

2.8   Key Expressions

Atomic mass expression:

Atomic mass = $\underline{\text{% abundance of isotope 1}}$ · mass of isotope 1 +
                             100

   $\underline{\text{% abundance of isotope 2}}$ · mass of isotope 2 + ...
                  100

Mole concept expression:

   1 mole = 6.022×10^{23} particles

2.9   Hints for Answering Questions and Solving Problems

Working with the mole concept:

To convert atoms to moles, use

   atoms · $\dfrac{1 \text{ mole}}{6.022 \times 10^{23} \text{ atoms}}$

To convert moles to atoms, use

   moles · $\dfrac{6.022 \times 10^{23} \text{ atoms}}{1 \text{ mole}}$

To convert grams to moles, use

   grams · $\dfrac{1}{\text{molar mass}}$

To convert moles to grams, use

moles · molar mass

To convert atoms to moles to grams, use

$$\text{atoms} \cdot \frac{1 \text{ mole}}{6.022 \times 10^{23} \text{ atoms}} \cdot \text{molar mass}$$

To convert grams to moles to atoms, use

$$\text{grams} \cdot \frac{1}{\text{molar mass}} \cdot \frac{6.022 \times 10^{23} \text{ atoms}}{1 \text{ mole}}$$

CHAPTER 3

MOLECULES AND COMPOUNDS

This chapter describes elements and compounds that are composed of molecules and compounds that are composed of charged atoms or groups of atoms called ions, ionic compounds. Guidelines are given for predicting the charges on monatomic ions and for writing the formulas of ionic compounds when the formulas for the ions can be predicted or are given. Systematic methods for naming both molecular and ionic compounds are described, as are mole concept problems involving chemical compounds and means of determining formulas for chemical compounds from data from chemical analyses. These concepts are central to an understanding of chemistry and should be mastered at this point in your study of chemistry.

3.1 Elements That Exist As Molecules

Most of the known elements are metals. Mercury is the only metal that exists as a liquid at room temperature and pressure, but gallium has such a low melting point that it melts when held in the hand.

Nonmetals can exist as solids, liquids or gases at room temperature and pressure.

- Hydrogen, nitrogen, oxygen, fluorine and chlorine exist as gases composed of H_2, N_2, O_2, F_2 and Cl_2 molecules, respectively,

- bromine exists as a liquid composed of Br_2 molecules,

- iodine exists as a solid composed of I_2 molecules,

- phosphorous exists as a solid composed of tetrahedral P_4 molecules and

- sulfur exists as a solid composed of crown-shaped S_8 molecules.

Oxygen, phosphorus and sulfur also exist in different elemental forms in the same physical state, forms called **allotropes**. The most familiar of these is ozone, O_3. Lastly, carbon can exist in the allotropic form of C_{60} ("buckyball") molecules, as well as

---

the extended networks of carbon atoms known as graphite and diamond. ***The properties of elements and compounds are related to their structures, as evidenced by the properties of graphite and diamond.*** ✷ Try preparing a "flash card" summarizing the forms of the elements that commonly exist as molecules and memorizing them. Try preparing a second "flash card" summarizing the structures and characteristics of graphite and diamond and memorizing them.

3.2   Molecular Compounds

Compounds are pure substances that can be decomposed into two or more different pure substances. Compounds composed of just nonmetals commonly exist as molecules. The subscripts in the **molecular formulas** for these molecular compounds give the exact numbers of atoms of each element in each molecule of the compound. The **structural formulas** for these compounds emphasize the connectivity of atoms and any chemically important groups of atoms in the molecules. Thus, the molecular formula for acetic acid, the component of vinegar that causes the tart taste, is $C_2H_4O_2$, and the structural formula is commonly written as CH_3COOH or CH_3CO_2H. The structural formula tells us that three atoms of H are connected (bonded) to one of the carbon atoms, which is itself bonded to the second carbon atom. In addition, one oxygen atom and one OH group are bonded to the second carbon atom giving the -COOH or $-CO_2H$ **functional group** characteristic of organic acids called carboxylic acids.

Molecular Models

The structures of molecular compounds can be represented by models showing the connectivity between atoms and the positions of atoms in space. Computer-generated models are now being used to design drugs, such as those being developed to combat aids, and to even map enormously large molecules, such as that of the common cold virus.

3.3   Ions

Atoms of almost all elements, other than those of Group 8, tend to gain or lose electrons in chemical reactions to form **ions**, which are atoms or groups of atoms bearing a net charge. Compounds composed of metals and nonmetals or groups of nonmetals are usually composed of ions and are called **ionic compounds**. A characteristic of metals is the tendency of their atoms to lose electrons and form positively charged ions called **cations** (cat'-ions). A characteristics of nonmetals is the tendency of their atoms to gain electrons to form negatively charged ions called **anions** (ann'-ions).

Charges on Monatomic Ions

The metals of Groups 1A-3A form cations having a positive charge equal to the group number of the metal. The transition metals typically form cations having 2+ and 3+ charges, and the nonmetals often form anions having a negative charge equal to 8 minus the group number of the nonmetallic element. ∗ Try preparing a "flash card" summarizing the charges formed by elements of the A Groups of the periodic table and memorizing them. This could involve using a verbal summary or a drawing of the form of Fig. 3.10 of the text showing the charges of ions as a function of the periodic positions of the elements.

Valence Electrons

The total number of electrons for monatomic ions formed by elements of the A Groups in the periodic table always equals the number of electrons of an atom of a noble gas. Atoms of metals lose sufficient electrons to achieve the same number of electrons as the nearest noble gas of lower atomic number whereas atoms of nonmetals gain sufficient electrons to achieve the same number of electrons as the nearest noble gas of higher atomic number. Thus, atoms of K and Ca lose one and two electrons, respectively, to achieve the same number of electrons as Ar whereas atoms of Se and Br gain two and one electrons, respectively, to achieve the same number of electrons as Kr.

G.N. Lewis first suggested electrons were arranged in shells about the nucleus and that only electrons in the outermost shell were involved when atoms combined with one another. These outermost electrons came to be known as **valence electrons**. The number of valence electrons possessed by atoms of elements of the A Groups of the periodic table is equal to the Group number.

Polyatomic Ions

A **polyatomic ion** is composed of two or more atoms that are bonded together and carry a net electrical charge. *The ammonium ion, NH_4^+, is the only common cation that is not a metal cation.*

3.4   Ionic Compounds

Cations and anions of opposite electrical charge are brought together by electrostatic forces of attraction to form ionic

---

*compounds. The cations and anions always combine in the simplest
ratio leading to electrical neutrality for the ionic compound,
and this observation can be used to predict the formulas of
ionic compounds. The symbol of the cation is always written
first, followed by the symbol of the anion. When a polyatomic
ion appears more than once in a formula, it is written in
parentheses and a subscript is used to the indicate the number
of polyatomic ions appearing in the formula.*

The Ionic Crystal Lattice

*Ionic compounds are generally solids that have their cations and
anions arranged in regular patterns in extended three-dimensional networks.* The regular array of positive and negative ions
is called a **crystal lattice**. The simplest combination of cations
and anions showing electrical neutrality is called a **formula
unit**. *A formula unit is merely a hypothetical unit, because no
discrete formula units exist in the crystal lattice.*

Properties of Ionic Compounds

*The strong electrostatic forces of attraction within crystal
lattices cause ionic compounds to be solids with high melting
points. Most ionic compounds do not conduct electricity as
solids because the ions are held in fixed positions in the
crystal lattice by the strong electrostatic forces of attraction. However, the crystal lattice is broken when an ionic
compound is melted, and the ions are free to move, so molten
(melted) ionic compounds are generally good conductors of
electricity.* Ionic compounds also produce ions by **dissociation**
when they dissolve in water; the interactions between the ions
and the oppositely charged ends of the water molecules contribute to the energy that is required to overcome the forces of
attraction between the cations and anions. Compounds whose
aqueous solutions conduct electricity are called **electrolytes**,
and all ionic compounds that are soluble in water are good
electrolytes. ✶ Try preparing a "flash card" stating the properties of ionic
compounds on one side and the reasons for these properties on the other side and
memorizing them.

3.5   Names of Compounds

Most compounds are named following rules developed by internati-
onal committees. Some, however, have common names like water and
ammonia.

Naming Ionic Compounds

The name of an ionic compound is based on the names of the positive and negative ions in the compound. ***The name of the positive ion, the cation, is given first and in most cases this corresponds to the name of the metal in the compound. In cases where the metal can have more than one positive charge, as with the transition and post-transition metals, the most common practice is to indicate the charge of the metal cation by placing a Roman numeral equal to the charge in parentheses immediately following the name of the metal.*** For example, copper(I) is used for Cu^+, iron(II) is used for Fe^{2+} and tin(IV) is used for Sn^{4+}. The only common nonmetal cation is NH_4^+.

Monatomic anions are named by adding -ide to the stem of the name of the nonmetallic element from which the anion is derived, as shown in Fig. 3.16 in the text. The names of polyatomic anions are given in Table 3.2 in the text. ✶ Try preparing a separate "flash card" for each separate group or area of the periodic table and memorizing the names of the polyatomic ions for that group or area.

Naming Binary Compounds of the Nonmetals

Compounds composed of two nonmetals usually exist as molecules. These compounds are also named in a systematic manner. We will begin with naming **binary compounds** of hydrogen, that is with two-element compounds containing hydrogen, and then consider the general pattern that is used for naming compounds composed of two nonmetallic elements from Groups 4A-7A. ***Consider the following***:

- When hydrogen is combined with oxygen, sulfur or the halogens, the H atom is generally written first and named first. The other element is named as though it were a negative ion, as with hydrogen fluoride for HF and dihydrogen selenide for H_2Se.

- When hydrogen is combined with carbon to form the **alkane** series of compounds having the general formula C_xH_{2x+2}, the first four members of the series are usually called the common names methane, ethane, propane and butane and those having 5 or more carbon atoms are named by using a prefix to indicate the number of carbon atoms and a suffix of "ane", as with octane for C_8H_{18}. ✶ Try preparing a "flash card" showing the common names and formulas for the first four alkanes and memorizing them. Also try preparing a "flash card" showing the general formula for alkanes and an example name and formula of an alkane having 5 or more carbon atoms and memorizing them, so you can use this information to write the names and formulas of other alkanes having 5 or more carbon atoms.

• When naming binary compounds composed of other nonmetallic elements from Groups 4A-7A, prefixes, such as "mon-, di-, tri-, tetra-, penta-, etc.", are used to indicate the number of atoms of each element in the compound, though the prefix "mon" is never used with the first element when just one atom of the first element occurs in the formula. Examples include: carbon monoxide for CO, sulfur trioxide for SO_3, dinitrogen monoxide for N_2O and tetraphosphorus decaoxide for P_4O_{10}. ***However, some binary compounds composed of nonmetals have common names that should be learned.*** ✶ Try preparing a "flash card" listing the formulas and common names of these binary compounds and memorizing them.

3.6 Molecules, Compounds and The Mole

The formula of a molecular element or of a compound can be used to calculate the **molar mass**, ***M***, of the substance. The molar mass of N_2O_4, for example, is equal to two times the molar mass of nitrogen plus four times the molar mass of oxygen, 92.02 g/mol. Chemists sometimes use the terms molecular weight for molecular elements and compounds and formula weight for ionic compounds in place of molar mass; these terms mean the same thing, but molar mass is more general and is therefore preferable.

The molar mass is the mass in grams of Avogadro's number of molecules for molecular elements and compounds and of Avogadro's number of formula units for ionic compounds. Thus, it is possible to determine the mass of one molecule or one formula unit in grams by simply dividing the molar mass by Avogadro's number. It also is possible to determine the mass of any given number of molecules or formula units or conversely the number of molecules or formula units in any given mass of substance using the mole concept. Furthermore, the chemical formula can be used to calculate the number of atoms of any element or the number of ions of any type present, once the number of molecules or formula units has been determined. These calculations are illustrated in Ex. 3.6 in the text.

3.7 Describing Chemical Formulas

The branch of chemistry known as chemical analysis is used to determine the formulas and structures of compounds. The compositions of compounds are often given in terms of the percentage by mass composition.

Mass Percent Composition

According to the law of constant composition any sample of a pure compound always consists of the same elements in the same proportion by mass. Thus, it is possible to express the composition of a compound in terms of the **percent composition** of each element in the compound by mass.

Percent always equals the part over the whole times 100%. Thus, the mass percent calcium in $Ca_3(PO_4)_2$ is given by

$$\%Ca = \frac{\text{mass of Ca in 1 mol } Ca_3(PO_4)_2}{\text{molar mass } Ca_3(PO_4)_2} \times 100\%$$

$$= \frac{3(\text{molar mass of Ca})}{\text{molar mass } Ca_3(PO_4)_2} \times 100\%$$

The mass percent of phosphorus is similarly given by

$$\%P = \frac{\text{mass of P in 1 mol } Ca_3(PO_4)_2}{\text{molar mass } Ca_3(PO_4)_2} \times 100\%$$

$$= \frac{2(\text{molar mass of P})}{\text{molar mass } Ca_3(PO_4)_2} \times 100\%$$

and that of oxygen by

$$\%O = \frac{\text{mass of O in 1 mol } Ca_3(PO_4)_2}{\text{molar mass } Ca_3(PO_4)_2} \times 100\%$$

$$= \frac{8(\text{molar mass of O})}{\text{molar mass } Ca_3(PO_4)_2} \times 100\%$$

The sums of the mass percents of the components of a compound must total 100%, so that the mass percent of oxygen also could be calculated by using 100% - %Ca - %P.

Empirical and Molecular Formulas

The mass percent composition of a compound can be used to determine the **empirical formula** of the compound, the formula showing the simplest ratio of moles of atoms of each component. *Since percent always refers to parts per hundred, the mass percent composition of a compound can be used to determine the number of grams of each element that would be present in exactly 100 grams of the compound. These masses can be converted to moles of atoms by multiplying by 1/molar mass, and the relative number of moles of atoms of each element can be determined by dividing each number of moles by the smallest number of moles of atoms present.* The latter procedure yields one mole of atoms of the element present in the least number of moles of atoms in the formula and gives all other moles of atoms relative to this element. *If whole numbers are not obtained, each number of moles is multiplied by integer numbers starting with two and proceeding until whole numbers of moles of atoms are obtained for each element.* This procedure is illustrated on page 137 and in Ex. 3.7 in the text.

The molecular formula shows the actual numbers of moles of atoms of each element present in one mole of the compound, rather than the relative numbers of moles of atoms of each element, and the subscripts in the molecular formula are always a whole number multiple of the subscripts in the corresponding empirical formula. To be able to determine the value of this whole-number multiple, it is necessary to know the molar mass of the compound from a separate experiment. *The whole-number multiple is then determined by dividing the molar mass of the compound by the simplest formula mass* as shown on page 137 of the text. In cases where the whole-number multiple is equal to one, the molecular formula is the same as the empirical formula, as it is in Ex. 3.7 in the text.

Determining and Using Formulas

In some cases, the combining masses of elements are determined. These masses can be converted directly to moles of elements to determine the empirical formula of the compound as illustrated on page 140 and in Ex. 3.8 of the text. However, the molar mass of the compound is still required to determine whether the empirical formula that is obtained in this manner is also the molecular formula for the compound. *If the compound is an ionic compound, the empirical formula is almost always the actual formula for the compound, because ionic compounds are represented by the simplest formulas showing electrical neutrality.*

Once the formula of a compound has been determined, it can be used in chemical calculations. Ex. 3.9 in the text illustrates the calculation of amount of compound that can be formed from a given mass of an element whereas Exercise 3.14 asks what mass of an element can be obtained from a given mass of a compound. *Notice that the mole concept is used in problems based on chemical formulas, because chemical formulas tell us the masses of elements present in one mole of compound as well as the mass of one mole of compound.* Chemical formulas can therefore be used to write useful conversion factors.

3.8 Hydrated Compounds

If ionic compounds are prepared in water solutions, in aqueous solutions, and isolated by evaporation, molecules of water can be associated with ions of the compound and be incorporated into

the lattice of the compound. Compounds containing chemically combined water in this manner are called **hydrated compounds** or simply **hydrates.** *There is no simple way to predict how much water will be incorporated into a hydrated compound; it must therefore be determined experimentally.* A sample of the hydrated compound is heated so all of the water is released from the compound and vaporized and only the **anhydrous** compound remains. The mass of water expelled equals the difference between the original mass of sample and the mass of the anhydrous residue. The masses of water and anhydrous compound are each converted to moles by multiplying by 1/molar mass, and the ratio of moles of water to moles of anhydrous compound is calculated to determine the formula of the hydrated compound. This procedure is illustrated in Ex. 3.10 in the text.

3.9 Key Expressions

Mass Percent Expression:

mass % of element = mass of element in 1 mol of compound x 100%
   in a compound   molar mass of compound

where mass of element in 1 mol of compound equals (number of moles of element in 1 mol of compound)(molar mass of element).

3.10 Hints for Answering Questions and Solving Problems

Writing formulas for ionic compounds given ions:

The formula for an ionic compound composed of C^{x+} cations and A^{y-} anions can be written by using the numerical charge of the anion as the subscript of the cation and the numerical charge of the cation as the subscript of the anion and reducing the subscripts to give the smallest whole numbers; that is by writing $C_y A_x$ and making sure that y and x are the smallest numbers giving electrical neutrality. In the case of Ca^{2+} and Br^- this process yields $CaBr_2$, because the numerical charge of Br^- is understood to be one and subscripts of one are not written and are also understood to be one. In the case of Al^{3+} and O^{2-} this process yields Al_2O_3, and in the case of Mg^{2+} and S^{2-} this process yields Mg_2S_2, which reduces to MgS.

Working with the mole concept:

To convert g → mol → molecules or formula units → atoms or ions use

$$g \cdot \frac{1}{M} \cdot \text{Avogadro's \#} \cdot \text{\# atoms per molecule or \# atoms or ions per formula unit}$$

To convert atoms or ions → molecules or formula units → mol→ g use

$$\text{atoms or ions} \cdot \frac{1}{\substack{\text{\# atoms per molecule or} \\ \text{\# atoms or \# ions per} \\ \text{formula unit}}} \cdot \frac{1}{\text{Avogadro's \#}} \cdot M$$

Consult Ex. 3.6 in the text for illustration of some of these conversions, and note that the logic maps that are indicated by the strings of arrows above can be entered or exited at any point.

Working with analytical data to determine empirical and molecular formulas:

Empirical formulas show the simplest ratios of moles of atoms of each element present in the compound. To calculate the numbers of moles of atoms of each element, use

$$\text{g of element} \cdot \frac{1}{\text{molar mass of element}}$$

for each element, with grams of element being obtained directly from the data given or from grams per 100 grams of sample when percent by mass data is given. Then divide each number of moles of atoms by the smallest number to obtain the ratios of moles of atoms in the compound. If this does not produce whole numbers, multiply by integer numbers starting with two until whole numbers are obtained.

Convert the empirical formula to a molecular formula by using

$$\text{whole-number multiple} = \frac{\text{molar mass of compound}}{\text{simplest formula mass in grams}}$$

to calculate the number that is to be used to multiply the subscripts in the simplest formula.

CHAPTER 4

PRINCIPLES OF REACTIVITY:
CHEMICAL REACTIONS

Chemical reactions are the key to life. We depend on thousands of chemical reactions taking place within our bodies and are affected by chemical reactions which take place daily in industry and in our environment. The goal of Chapter 4 is to introduce you to the concept of using balanced chemical equations to represent chemical reactions, to introduce you to several common types of chemical reactions, to teach you how to recognize and predict the outcome of these types of reactions and to teach you how to write balanced overall and net ionic equations for these types of reactions.

4.1   Chemical Equations

Chemical reactions are represented by **chemical equations** showing the formulas and physical states of the substances involved. The substances combined in the reaction are called **reactants** and are shown on the left side of the equation. The substances produced in the reaction are called **products** and are shown on the right side of the equation. The reactants and products are separated by an arrow, which can be interpreted to mean "produces" or "yields." The general form of a chemical equation is: **reactants → products.** For example,

$$2HCl(aq) + Na_2CO_3(s) \rightarrow 2NaCl(aq) + H_2O(\ell) + CO_2(g)$$

represents the reaction of hydrochloric acid with solid sodium carbonate to produce sodium chloride in the aqueous (water-based) solution, liquid water and gaseous carbon dioxide.

Chemical equations must be balanced to show equal numbers of each kind of atom on each side of the equation, because matter is neither created nor destroyed during chemical reactions. Equations are balanced by placing multiplying coefficients in front of the formulas for the reactants and products. (The absence of a coefficient indicates a coefficient of one.) The atom balance for the example equation given above is, therefore,

	2 atoms H,	→	2 atoms H,
	2 atoms Cl,		2 atoms Cl,
	2 atoms Na,		2 atoms Na,
	1 atom C, and		1 atom C, and
	3 atoms O.		3 atoms O.

A balanced equation also gives the mass relations between the various reactants and products, so the coefficients in a balanced equation are called **stoichiometric coefficients.**

4.2   Balancing Chemical Equations

A chemical equation must contain the correct chemical formulas for reactants and products. Thus subscripts in the formulas of the reactants and products can never be changed to balance equations. Instead multiplying numbers called stoichiometric coefficients are placed in front of the formulas, and *equations usually are balanced using the smallest set of whole-number coefficients*.

The general approach to balancing equations involves changing coefficients one at a time until the number of atoms of each element is the same on both sides of the equation. It is usually best to start by balancing those elements that appear in only one substance on each side of the equation. For example, to balance $C_6H_{14}(\ell) + O_2(g) \rightarrow CO_2(g) + H_2O(\ell)$, we should start by balancing C or H, but not O, and proceed to change the coefficients of the other substances as necessary to achieve an atom balance. *In the case of reactions involving polyatomic ions that remain intact,* such as $Al(s) + CuSO_4(aq) \rightarrow Al_2(SO_4)_3(aq) + Cu(s)$, *we should start by balancing the number of polyatomic ions on each side.* In this case, we should place a 3 in front of the $CuSO_4$ to balance the SO_4^{2-} ions and proceed to place a 2 in front of the Al and 3 in front of the Cu to obtain: $2Al(s) + 3CuSO_4(aq) \rightarrow Al_2(SO_4)_3(aq) + 3Cu(s)$.

The process of balancing equations is often one of trial and error. There will be cases where changing a coefficient to balance one element will unbalance another element or elements and require you to go back and change other coefficients to rebalance atoms that have become unbalanced. Thus, if at first you don't succeed, try again. However, following the hints given above will make the task easier.

Types of Reactions: Combustion Reactions

Combustion reactions commonly involve reaction with oxygen. The products of complete combustion of compounds containing only C and H (hydrocarbons) and of compounds containing only C, H, and O are always carbon dioxide and water. Incomplete combustion yields carbon monoxide, or a mixture of carbon monoxide and carbon dioxide, and water.

The balanced equation for the complete combustion of ethyl alcohol, CH_3CH_2OH, the alcohol that is present in alcoholic beverages, is: $2CH_3CH_2OH(\ell) + 6O_2(g) \rightarrow 4CO_2(g) + 6H_2O(\ell)$. The energy that is liberated as the combustion of ethyl alcohol takes place in our bodies is the source of "nutritional" calories, and ethyl alcohol is said to be high in calories.

4.3   Properties of Compounds in Aqueous Solution

Ions in Aqueous Solution

Solutions for which water is the solvent are called **aqueous solutions**. Substances which produce ions in aqueous solutions and therefore give solutions that conduct electricity are called **electrolytes**. Ionic compounds break apart to produce ions by **dissociation** in water whereas those molecular compounds that do produce ions in water do so in **ionization** reactions with water. Substances which do not produce ions in aqueous solutions are called **nonelectrolytes.**

Electrolytes can be classified as *strong* or *weak* on the basis of the conductivity of the aqueous solutions. ***Strong acids and bases and ionic salts are strong electrolytes. Weak acids and bases are weak electrolytes***, because they do not react with water to produce appreciable concentrations of ions, do not **ionize** appreciably in water. Acids and bases are discussed in Section 4.4, and a list of common strong and weak acids and bases is given in Table 4.1 in the text. ✳ Try preparing a "flash card" listing these names and formulas and memorizing them.

Nonelectrolytes are generally molecular compounds which do not react with water. Examples include methyl alcohol (CH_3OH), ethyl alcohol (CH_3CH_2OH), urea (($NH_2)_2CO$), sugar as sucrose ($C_{12}H_{22}O_{11}$), starch and antifreeze (ethylene glycol, HOC_2H_4OH). ✳ Try preparing a "flash card" listing these names and formulas and memorizing them.

Solubility of Ionic Compounds in Water:

Ionic compounds dissolve in water to varying extents. Some are essentially insoluble, others are moderately soluble and still others are very soluble. **You can assume that any ionic compound that contains at least one of the ions listed in the top portion of Fig. 4.7 in the text will be at least moderately soluble in water**. ✳ Try preparing a "flash card" listing the names and formulas of ions forming soluble salts and another "flash card" listing names and formulas of ions forming insoluble salts and memorizing them.

4.4 Acids and Bases

An **acid** is any substance that increases the concentration of hydrogen, H^+, ions when dissolved in pure water. A number of general characteristics of acids are given in the introduction to this Section in the text. ✴ Try preparing a "flash card" listing these general properties of acids and memorizing them.

A **base** is any substance that increases the concentration of hydroxide, OH^-, ions when dissolved in pure water. Several characteristics of bases are given in the introduction to this Section in the text. In addition, you should know that a base can **neutralize** the effect of an acid, and an acid can neutralize the effect of a base. ✴ Try preparing a "flash card" listing these general properties of bases and memorizing them.

Types of Reactions: Behaviors of Substances as Acids and Bases in Water

Strong acids are completely converted to ions in ionization reactions with water whereas **weak acids** are only partially converted to ions in ionization reactions with water; the percentage ionization for weak acids is often less than 10%.

Strong acid: $HA(aq) + H_2O(\ell) \rightarrow H_3O^+(aq) + A^-(aq)$
                    $\approx 100\% \text{ reaction}$
or simply: $HA(aq) \rightarrow H^+(aq) + A^-(aq)$
                    $\approx 100\% \text{ reaction}$

Weak acid : $HA(aq) + H_2O(\ell) \rightarrow H_3O^+(aq) + A^-(aq)$
                    $\approx <10\% \text{ reaction}$
or simply: $HA(aq) \rightarrow H^+(aq) + A^-(aq)$
                    $\approx <10\% \text{ reaction}$

Strong **bases** are usually ionic metal hydroxides which completely dissociate into ions in water whereas **weak bases** are usually molecular compounds which are only partially converted to ions in ionization reactions with water.

Strong base: $MOH(s) \xrightarrow{H_2O} M^+(aq) + OH^-(aq)$
                    $\approx 100\% \text{ reaction}$

or: $M(OH)_2(s) \xrightarrow{H_2O} M^{2+}(aq) + 2OH^-(aq)$
                    $\approx 100\% \text{ reaction}$

Weak base: $B(aq) + H_2O(\ell) \rightarrow HB^+(aq) + OH^-(aq)$

$\approx$ <10% reaction

The reactions that are listed as "$\approx$ 100% reactions" are for dilute solutions of the strong acids and bases listed in Table 4.1 in the text, solutions having concentrations comparable to those commonly encountered in the laboratory. However, in the case of the $M(OH)_2$ hydroxides from Group IIA that are commonly considered to be strong bases, $Ca(OH)_2$, $Sr(OH)_2$ and $Ba(OH)_2$, the solutions must be <0.013 M for $Ca(OH)_2$, <0.043 M for $Sr(OH)_2$ and <0.11 M for $Ba(OH)_2$, because these concentrations correspond to the maximum solubilities of these compounds. $Mg(OH)_2$ having a maximum solubility of only 0.00016 M is too insoluble to be considered to be a strong base.

In Chapter 17, you will learn that cations that differ from weak bases by H^+, such as NH_4^+ from NH_3, and aquometal cations, such as $Al(H_2O)_6^{3+}$, also act as weak acids. You also will learn that anions that differ from weak acids by H^+, such as $CH_3CO_2^-$ from CH_3CO_2H, act as weak bases. However, you will not need to be concerned with these types of weak acids and weak bases until you are studying Chapter 17.

Types of Reactions: Reactions of Metal and Nonmetal Oxides With Water

Oxides of nonmetals can react with water to produce acidic solutions, though they do not contain H atoms. For example, CO_2 produces H_2CO_3, SO_2 is converted to SO_3 which produces H_2SO_4 and NO_2 produces HNO_3. Each of these reactions is involved in the formation of acid rain. Nonmetal oxides which can react with water to produce H^+ ions are called **acidic oxides**. By way of contrast, metal oxides, which can dissolve in water, react with water to produce OH^- ions and are called **basic oxides**.

4.5 Equations for Reactions in Aqueous Solution: Net Ionic Equations

Ions that are present in solution but do not participate in the reaction of interest are called **spectator ions**. The balanced equation that results from leaving out the spectator ions shows only the species participating in the reaction and is called the **net ionic equation** for the reaction. ***There must always be a conservation of charge as well as mass, so balanced net ionic equations must meet these requirements.***

4.6   Types of Reactions in Aqueous Solution

Types of Reactions: Precipitation Reactions

A reaction in which ions combine to form an insoluble reaction product when water soluble reactants are mixed is called a **precipitation reaction**, and the insoluble product is called a **precipitate**. The general form of the overall equation for the formation of a precipitate is: **AB + CD → AD + CB** where one of the products, AD or CB, is insoluble. Reactions of this type also are known as **exchange or double displacement** reactions, since A and C exchange places and thus displace each other. The driving force for precipitation reactions is the formation of an insoluble substance.

The procedure for writing net ionic equations for precipitation reactions is illustrated in Ex. 4.3 in the text. The process can be summarized as follows:

> • Write a balanced overall equation for the reaction using complete formulas and indicating whether the substances are present in aqueous solution or present as pure solids, liquids or gases.

> • Consult solubility guidelines, such as Fig. 4.7 in the text, and rewrite the balanced overall equation in terms of ions for the water soluble substances and in terms of a complete formula for the insoluble product, the precipitate. This equation is known as a complete ionic equation.

> • Write a net ionic equation showing only the ions that combine to form the precipitate and the precipitate. This will require eliminating ions that appear in equal numbers on both sides of the complete ionic equation and do not participate in the reaction, ions that are spectator ions.

> • Check to make sure that your net ionic equation has both an atom and electrical charge balance and is written in terms of the smallest set of whole number coefficients.

* Try using these steps until they become so familiar that you can use them without taking too long to think about them.

Suppose solutions of water soluble AB and CD are mixed to give the reaction of the general overall equation, and insoluble AD and soluble CB are formed. The appropriate equations for the reaction would be:

Overall equation: AB(aq) + CD(aq) → AD(s) + CB(aq)

Complete ionic
equation: $A^+(aq) + B^-(aq) + C^+(aq) + D^-(aq) → AD(s) +$
                          $C^+(aq) + B^-(aq)$

Net ionic equation: $A^+(aq) + D^-(aq) → AD(s)$

This means A^+(aq) combines with D^-(aq) to form AD(s) while C^+(aq)
and B^-(aq) remain in solution as spectator ions. *The net ionic*
equations for precipitation reactions always involve the
combination of ions of opposite charges in ratios leading to
electrical neutrality for the precipitate.

Types of Reactions: Strong Acid – Strong Base Reactions

The general overall equation for a strong acid having one
ionizable H atom, HA, and a strong base metal hydroxide from
Group IA is: **HA(aq) + MOH(aq) → MA(aq) + $H_2O(\ell)$**. *The products of*
an acid-base reaction are usually a salt plus water. Notice,
also that this reaction is another example of an exchange or
double displacement reaction, since H^+ and M^+ exchange places and
thus displace each other.

The net ionic equation for the reaction of a strong acid with a
strong base is: $H^+(aq) + OH^-(aq) → H_2O(\ell)$, because both HA(aq)
and MOH(aq) exist as ions in solution and M^+(aq) and A^-(aq) are
spectator ions which do not participate in the reaction. Indeed,
this is the net ionic equation for the reaction of all strong
acids and strong bases, since all strong acids and strong bases
exist as ions in solution. The driving force for these reactions
is the formation of the weak electrolyte H_2O.

Types of Reactions: Gas-Forming Reactions

Metal carbonates react with acids forming carbonic acid,
$H_2CO_3(aq)$, which decomposes to $H_2O(\ell)$ and $CO_2(g)$. The form of the
net ionic equation for these reactions depends on whether the
metal carbonate is water soluble. In the case of water soluble
K_2CO_3, it is: $CO_3^{2-}(aq) + 2H^+(aq) → CO_2(g) + H_2O(\ell)$. In the case
of water insoluble $CaCO_3$(s), which is limestone, it is: $CaCO_3$(s)
+ $2H^+(aq) → Ca^{2+}(aq) + H_2O(\ell) + CO_2(g)$. This reaction is occurring
with increasing frequency as acid rain acts on limestone/marble
buildings and statutes throughout the world. *The driving force*
for these reactions is the formation of a water-insoluble gas,
CO_2.

Oxidation-reduction reactions involve the transfer of electrons from one substance to another. There is no general overall or net ionic equation that can be written for oxidation-reduction reactions. *The driving force for oxidation-reduction reactions is the transfer of electrons.* You will learn more about balancing equations for oxidation-reduction reactions when studying Chapter 21.

✳ Try referring to the summary given in the text to prepare a "flash card" summarizing these types of reactions and the driving forces for these types of reactions and memorize them. In addition, recall that the net ionic equation for the reaction of all strong acids and strong bases is: $H^+(aq) + OH^-(aq) \rightarrow H_2O(\ell)$.

4.7 Precipitation Reactions

A precipitation reaction produces an insoluble product, a precipitate. The mixing of two water soluble compounds releases the ions of these compounds to allow two new combinations of ions to form. If either one of these combinations leads to an insoluble compound, precipitation occurs. The solubility guidelines of Fig. 4.7 in the text can be used to determine whether an insoluble compound will form. Then the procedure for writing net ionic equations for precipitation reactions that was summarized in 4.6 can be followed to write the net ionic equation for the precipitation reaction.

4.8 Acid-Base Reactions

Acid-base reactions involving strong or weak acids reacting with strong bases produce a salt of the acid and water. Acid-base reactions involving strong or weak acids reacting with weak bases produce only salts (Ex. 4.5 in the text). Thus, the word "**salt**" is commonly used to refer to any ionic compound whose cation comes from a base and whose anion comes from an acid and is not restricted to only referring to common table salt, NaCl.

> Note: The category "salts" does not include the ionic metal oxides and hydroxides, because these compounds are usually classified as bases.

The net ionic equation for the reactions of all strong acids with all strong bases is: $H^+(aq) + OH^-(aq) \rightarrow H_2O(\ell)$. This means the complete reactions of strong acids with strong bases result in **neutral solutions**, solutions that are neither acidic nor basic. Thus acid-base reactions are generally called **neutralization reactions.** However, when studying Chapter 18, you will learn that so-called "neutralization" reactions involving either weak acids or weak bases or both seldom yield neutral solutions.

4.9   Gas-Forming Reactions

The reactions of metal carbonates with acids to produce carbonic acid, $H_2CO_3(aq)$, which subsequently decomposes to $H_2O(\ell)$ and $CO_2(g)$ were discussed in 4.6. Metal hydrogen carbonates (bicarbonates) also react with acids to produce $CO_2(g)$. In addition, metal sulfites react with acids to produce pungent $SO_2(g)$, and metal sulfides react with acids to produce $H_2S(g)$, which smells like rotten eggs. The smell of rotten eggs actually comes from $H_2S(g)$ that is formed by bacterial decomposition of sulfur-containing proteins. Furthermore, both $SO_2(g)$ and $H_2S(g)$ are quite toxic.

Gas formation reactions are also examples of exchange or double displacement reactions. The metal of the metal carbonate or metal hydrogen carbonate and the hydrogen of the acid exchange places producing $H_2CO_3(aq)$, which subsequently decomposes to $H_2O(\ell)$ and $CO_2(g)$; the metal of the metal sulfite and the hydrogen of the acid exchange places producing $H_2SO_3(aq)$, which decomposes to $H_2O(\ell)$ and $SO_2(g)$; and the metal of the metal sulfide and the hydrogen of the acid exchange places producing $H_2S(g)$.

4.10   Oxidation-Reduction Reactions

The word "oxidation" was originally used to refer to the reaction of an element with oxygen to form an oxide, the gain of oxygen. The word "reduction" was originally used to refer to the removal of oxygen from an oxide to form the element, the reduction of oxygen. Thus the rusting of iron metal to form Fe_2O_3 is an example of the oxidation of iron whereas the conversion of Fe_2O_3 in iron ore to iron metal is an example of the reduction of iron. Indeed, any reaction in which oxygen is added to another substance is **oxidation**, and the substance causing the oxidation is the **oxidizing agent**. Any reaction in which oxygen is removed from a substance is **reduction**, and the substance causing the reduction is the **reducing agent**.

The examples used in the text illustrate three important points concerning oxidation-reduction reactions:

• Oxidation is the opposite of reduction; the addition of oxygen is oxidation, and the removal of oxygen is reduction.

• The reducing agent reduces another substance and is itself oxidized, and the oxidizing agent oxidizes another substance and is itself reduced.

• If one substance is oxidized in a chemical reaction, another must be reduced. This is why these reactions are called oxidation-reduction reactions, or **redox reactions**, for short.

Not all redox reactions involve oxygen. However, **all redox reactions do appear to involve the transfer of electrons between substances.** When a substance loses electrons, it is said to be **oxidized.** When a substance gains electrons, it is said to be **reduced**. A helpful device for remembering this is: Oil Rig. **Oil rig stands for: _Oxidation is loss_ of electrons and _reduction is gain_ of electrons.** No matter how you remember these terms, it is important that you realize that: **In every oxidation-reduction reaction, one reactant is reduced (and is therefore the oxidizing agent) and one reactant is oxidized (and is therefore the reducing agent).**

The key to identifying an oxidation-reduction reaction and determining which substance has gained electrons and is the oxidizing agent and which substance has lost electrons and is the reducing agent is to look for a change in the oxidation number of an element. The **oxidation number** of an atom in a molecule or ion is defined as the electrical charge an atom has or *appears* to have based on following some general guidelines for assigning oxidation numbers. It is important for you to realize that **oxidation numbers only correspond to actual electrical charges in the cases of monatomic ions.** However, oxidation numbers provide another means of describing oxidation and reduction and thereby provide a way of identifying oxidation-reduction reactions.

"Guidelines for Determining Oxidation Numbers" are given in the text. Keep in mind that **the oxidation number of an atom of a pure element is zero, no matter how complex the formula for the elemental form.** This means the oxidation numbers of the atoms in Na, O_2, P_4, S_8 and C_{60} are all zero. Also keep in mind that **the algebraic sum of the oxidation numbers for any compound must equal zero and that for any ion must equal the electrical charge on the ion.** These concepts are illustrated in Ex. 4.8 in the text. ✷ Try preparing a "flash card" summarizing the guidelines for determining oxidation numbers and memorizing them. Also try using them until they become so familiar that you can use them without taking too long to think about them.

The examples used in the text illustrate several important points concerning the relationship between oxidation numbers and oxidation-reduction concepts:

• Oxidation involves an increase in the oxidation number of an atom whereas reduction involves a decrease (reduction) in the oxidation number of an atom.

• Oxidizing agents contain elements which undergo a decrease in oxidation number and therefore are reduced whereas reducing agents contain elements which undergo an increase in oxidation number and therefore are oxidized. For example, NO_3^- acts as an oxidizing agent in its reaction with Cu in acidic solution, but only the N atom is reduced, because its oxidation number changes from +5 in NO_3^- to +4 in NO_2.

Thus, there are three ways of describing oxidation and reduction:

• in terms of gain and loss of oxygen,

• in terms of loss and gain of electrons and

• in terms of increase and decrease in oxidation number, respectively.

These three ways are described in greater detail in Table 4.3 of the text. ✷ Try preparing a "flash card" containing this information and memorizing it.

The most general and most useful of the three methods of describing oxidation-reduction reactions is actually in terms of oxidation numbers, though the most common method is in terms of loss and gain of electrons. Consider the reaction $PCl_3 + Cl_2 \rightarrow PCl_5$. There is no gain or loss of oxygen, and furthermore no gain or loss of electrons, because all of the substances are covalent. However, the oxidation number of P increases from +3 to +5 and the oxidation number of each atom of Cl in Cl_2 changes from 0 to -1, and the reaction is classified as an oxidation-reduction reaction.

The formulas of some common oxidizing and reducing agents are given in Table 4.2 in the text. *Notice that the common oxidizing agents are nonmetallic elements in elemental form or species containing elements in high oxidation states, since these elements are readily reduced. Also notice the common reducing agents are elements that are reasonably easy to oxidize, namely carbon, hydrogen and the metals.* *Try preparing a "flash card" listing these substances and ions, as well as the products of their actions, and memorizing them.

4.11 Hints for Answering Questions and Solving Problems

Writing Balanced Overall Equations:

Several hints for balancing overall chemical equations were given in 4.2. *First, you should start by balancing elements which occur in only one reactant and only one product. Second, if polyatomic ions are involved and remain intact, you should start by balancing the number of polyatomic ions of a given type.* In either case, you should proceed to change other coefficients by trial and error until you obtain the same number of atoms of each element on both sides of the equation. *You may use fractional coefficients to facilitate the balancing process, but the final overall balanced equation should be written using the smallest set of whole-number coefficients.*

Furthermore, *the key to completing and balancing equations for the complete neutralization reactions of strong and weak acids with strong bases is to recall that the products of these reactions are a salt and water and to realize that the goal is to react equal numbers of H^+ and OH^- ions.* Consider the reaction of H_3PO_4 with $Ca(OH)_2$: $H_3PO_4(aq) + Ca(OH)_2(aq) \rightarrow ?$. There are three H atoms in H_3PO_4 and two OH^- ions in $Ca(OH)_2$. The lowest common multiple of three and two is six, so these substances will react in the ratio $2H_3PO_4(aq) + 3Ca(OH)_2(aq) \rightarrow ?$. The salt that will be formed will be $Ca_3(PO_4)_2$, and there are several ways of predicting this. First, the formula is suggested by the combining ratio that was obtained in attempting to react equal numbers of H^+ and OH^- ions. Second, the charge for the calcium ion is predicted to be 2^+ based on its position in the Periodic Table, and that for the phosphate ion is predicted to be 3^- based on removal of three H^+ ions from H_3PO_4. The simplest ratio leading to electrical neutrality is therefore $Ca_3(PO_4)_2$, and the balanced overall equation is $2H_3PO_4(aq) + 3Ca(OH)_2(aq) \rightarrow Ca_3(PO_4)_2(s) + 6H_2O(\ell)$. The 6 H^+ and 6 OH^- ions combine to form 6 H_2O, and the reaction happens to involve both acid-base neutralization and precipitation. Thus, *it is always a good idea to check the solubility guidelines of Fig. 4.7 in the text to determine whether precipitation can occur when the acid-base reaction forms the salt.*

Writing Balanced Net Ionic Equations:

The procedure for writing net ionic equations for precipitation reactions was summarized in 4.6, and the net ionic equation for reactions of strong acids with strong bases, namely $H^+(aq) + OH^-(aq) \rightarrow H_2O(\ell)$, was given in 4.6. **The key to writing complete ionic equations and proceeding to write net ionic equations is to:**

* use <u>ions</u> to represent strong acids, strong bases and soluble salts,

* use <u>complete formulas</u> to represent weak acids, weak bases, water, nonelectrolytes and insoluble salts and

* eliminate any ions that appear in equal numbers on both sides of the equation and do not participate in the reaction.

* Try preparing two separate "flash cards" listing types of substances to be represented as ions and types of substances to be represented as complete formulas in net ionic equations and memorizing them.

Identifying the Type of Reaction:

The following types of chemical reactions were introduced in this chapter: combustion, behaviors of substances as acids and bases in water, reactions of metal and nonmetal oxides with water, precipitation, acid-base, gas-forming and oxidation-reduction. **If the reaction:**

* **involves reaction with O_2 and produces $CO(g)$ or $CO_2(g)$ and $H_2O(g)$ or $H_2O(\ell)$**, it is both a combustion reaction and an oxidation-reduction reaction, since all combustion reactions are also oxidation-reduction reactions.

* **involves double displacement and results in the formation of an insoluble product,** it is a precipitation reaction, and the net ionic equation involves bringing the appropriate aqueous ions of opposite charges together in the smallest ratio leading to electrical neutrality for the precipitate.

* **involves double displacement and results in the formation of a salt and water,** it is an acid-base reaction, and the net ionic equation is:

- $H^+(aq) + OH^-(aq) \rightarrow H_2O(\ell)$ for reactions of strong acids with strong bases,

- $HA(aq) + OH^-(aq) \rightarrow A^-(aq) + H_2O(\ell)$ for reactions of weak acids with strong bases.

- **involves reaction of a metal carbonate, metal hydrogen carbonate, metal sulfite or metal sulfide and an acid,** it is a gas-forming reaction.

- **involves changes in oxidation numbers,** it is an oxidation-reduction reaction.

  - Examples include, but are not limited to, all reactions in which elements are converted to compounds or compounds are converted to elements, especially those involving single displacements following the form $A + BC \rightarrow AC + B$.

✱ Try preparing a separate "flash card" for each of these five types of reactions, listing the type of reaction on one side and the corresponding identifying characteristic(s) on the other side, and memorizing them. Also try preparing a "flash card" giving the general forms of the net ionic equations for the reactions of strong and weak acids with strong bases.

CHAPTER 5

STOICHIOMETRY

The Law of Conservation of Mass can be explained by the fact that atoms are neither created nor destroyed during chemical reactions; rather atoms are merely rearranged during chemical reactions. This is why we use atom balances to write balanced equations for chemical reactions. *The coefficients that are obtained in writing a balanced overall equation tell us the quantities of the substances involved in the reaction and therefore tell us the mass relations between all the substances in the reaction.* This information is necessary to conduct any chemical reaction. For example, suppose there is a train wreck and a tank car containing concentrated sulfuric acid is ruptured. It would be desirable to know how much sodium carbonate to bring to neutralize the sulfuric acid in a CO_2 gas-forming reaction like the one discussed in 4.6, and how much sodium sulfate will be formed. The goal of Chapter 5 is to teach you how to do these types of calculations for chemical reactions and how chemical reactions can be used for quantitative analyses.

5.1 Weight Relations in Chemical Reactions: Stoichiometry

The study of the quantitative nature of chemical formulas and chemical reactions is known as the study of **stoichiometry** (stoy-key-**ahm**-uh-tree; from the Greek words *stoicheion*, "element", and *metron*, "measure"). The coefficients in a balanced chemical equation are called **stoichiometric coefficients,** because they can be used to determine relations between the masses of reactants and products of chemical reactions.

The coefficients of a balanced chemical equation can be interpreted in terms of atoms, molecules or formula units and therefore moles, since there are 6.022×10^{23} of anything in one mole. The number of moles can then be converted to grams giving the mass relations for the reactants and products. Consider the development of these concepts using the following example:

4 Al(s)	+	3 O_2(g)	→	2 Al_2O_3(s)
4 atoms		3 molecules	→	2 formula units
$4(6.022 \times 10^{23})$ atoms		$3(6.022 \times 10^{23})$ molecules	→	$2(6.022 \times 10^{23})$ formula units
4 moles		3 moles	→	2 moles
108.0 g		96.0 g	→	204.0 g

There are several additional points that should be made in conjunction with this information:

- aluminum and oxygen always react in the same mole proportions to form aluminum oxide and therefore always *react in the same mass proportions, though the actual numbers of grams involved can vary*, and

- there can be a conservation of mass during a chemical reaction without having a conservation of moles.

Suppose we wanted to know the number of moles and number of grams of Al_2O_3 that could be prepared from 6.8 moles of Al. We could convert moles of Al to the stoichiometrically equivalent number of moles of Al_2O_3 and then proceed to convert moles of Al_2O_3 to grams of Al_2O_3 following the procedure introduced in Chapter 3. That is, we could follow the ***logic pattern***: *mol Al → mol Al_2O_3 → g Al_2O_3*.

The conversion of moles of Al to the stoichiometrically equivalent number of moles of Al_2O_3 requires the use of a stoichiometric factor. A **stoichiometric factor** is a mole-ratio factor relating moles of one substance appearing in a balanced equation to moles of another substance appearing in the same balanced equation. ***The stoichiometric factor comes directly from the coefficients in the balanced equation.*** This is why it is so important for you to be able to write correctly balanced equations for chemical reactions.

In the case at hand, the desired stoichiometric factor for converting moles of Al to the stoichiometrically equivalent number of moles of Al_2O_3 is obtained from the coefficients in the balanced equation 4 Al(s) + 3 O_2(g) → 2 Al_2O_3(s) and is 2 mol Al_2O_3/4 mol Al. Thus converting 6.8 moles of Al to moles of Al_2O_3 can be accomplished using:

$$6.8 \text{ mol Al} \cdot \frac{2 \text{ mol } Al_2O_3}{4 \text{ mol Al}} = 3.4 \text{ mol } Al_2O_3$$

and converting 6.8 mol Al to grams of Al_2O_3 can be accomplished by using:

$$6.8 \text{ mol Al} \cdot \frac{2 \text{ mol } Al_2O_3}{4 \text{ mol Al}} \cdot \frac{102.0 \text{ g } Al_2O_3}{1 \text{ mol } Al_2O_3} = 3.5 \times 10^2 \text{ g } Al_2O_3$$

These conversions are examples of mole-to-mole and mole-to-mass conversions for chemical reactions.

In other cases, it might be desirable to calculate the number of grams of Al_2O_3 that can be produced from a given number of grams of Al or the number of grams of O_2 required to react with a given number of grams of Al. The *logic patterns* for these **mass-to-mass conversions** would be *g Al → mol Al → mol Al_2O_3 → g Al_2O_3* and *g Al → mol Al → mol O_2 → g O_2*, respectively. Conversions of this type are illustrated in Ex. 5.1 in the text.

The general logic pattern for all mass-to-mass conversions for chemical reactions is:

$$g\ A \to mol\ A \to mol\ B \to g\ B \qquad (5a)$$

where A is the substance for which information is given and B is the substance for which information is requested. However, it should be emphasized that one can enter the logic pattern at any given point and exit the logic pattern at any given point. For example, the *mol Al → mol Al_2O_3* and *mol Al → g Al_2O_3* conversions shown above are examples of *mol A → mol B* and *mol A → g B* conversions, respectively.

The mathematical operations for carrying out the g A → g B conversions of the general logic pattern are as follows:

$$g\ A \cdot \frac{1}{M_A} \cdot \frac{y\ mol\ B}{x\ mol\ A} \cdot M_B \qquad (5b)$$

where M_A and M_B represent the molar masses of A and B and x and y represent the coefficients of A and B in the balanced chemical equation, the stoichiometric coefficients of A and B. In other words, we use

$$g\ A \cdot \frac{1\ mol\ A}{M_A'\ g\ A} \cdot \frac{y\ mol\ B}{x\ mol\ A} \cdot \frac{M_B'\ g\ B}{1\ mol\ B}$$

where M_A' and M_B' represent only the numerical portions of the molar masses of A and B. The advantage of using this method of representation is that it more clearly shows the unit conversions resulting in g B; the results are necessarily the same as those that are obtained by using the preceding expression. See the discussion of this material in PROBLEM SOLVING TIPS AND IDEAS 5.1 in the text.

5.2   Reactions In Which One Reactant Is Present in Limited Supply

Reactants are seldom combined in stoichiometric amounts to conduct chemical reactions. Often a large excess of a relatively inexpensive reactant is supplied to insure that a more expensive reactant will be completely converted to products. In these cases, **the starting amounts of two or more reactants are known, and it is our job to calculate the maximum amount of product** *that can be produced.* The compound that is completely consumed first is called the **limiting reactant**, because its amount determines or limits the amount of product that can be formed. The situation is similar to determining how many three-wheeled bicycles can be constructed from 8 frames and 34 wheels. The construction of 8 bicycles would require 8 frames and 24 wheels, so the limiting "reactant" would be the frames, the maximum number of bicycles that could be constructed would be 8 and there would be 10 (34-24) wheels left over. Notice how this example indicates that *the maximum amount of product is based on the limiting reactant.*

The general procedure for determining which reactant is the limiting reactant for x A + y B → products when masses of both A and B are given is to:

- calculate the number of moles of A and B available by converting g A to moles A and g B to moles of B,

- calculate the value of the ratio:

  moles of A available
  moles of B available

  by dividing moles of A by moles of B,

- calculate the value of the ratio:

  moles of A required
  moles of B required

  by dividing the coefficient x of the balanced equation by the coefficient y,

- compare the values of the ratios

  moles of A available   to moles of A required,
  moles of B available   moles of B required

  and determine that A is the limiting reactant when

  moles of A available   <   moles of A required
  moles of B available   moles of B required

  because there is too much B available for A, and that B is the limiting reactant when

  moles of A available   >   moles of A required
  moles of B available   moles of B required

because there is too much A available for B.

This procedure is also summarized in PROBLEM SOLVING TIPS AND IDEAS 5.3 in the text.

The maximum amount of product that can be formed is based on the limiting reactant, just as the number of bicycles that could be constructed was based on the limiting number of frames. **Thus the maximum amount of product that can be formed in a limiting reactant problem in which the masses of two or more reactants are given can be calculated by doing a mass-to-mass conversion starting with the grams of the limiting reactant.** This is illustrated in Ex. 5.2 in the text.

In the case of x A + y B → z C, the amount of excess reactant calculated to remain is given by (g A originally present + g B originally present - g C calculated to be formed), because of the Conservation of Mass. However, in more complicated cases, it is necessary to:

- use the moles of limiting reactant and the appropriate stoichiometric factor to calculate the moles of "excess" reactant consumed,

- use (original moles of excess reactant - moles of excess reactant consumed) to calculate moles of excess reactant remaining, and

- convert moles of excess reactant remaining to grams of excess reactant remaining.

Notice how the Conservation of Mass could have been used to calculate that 14 g H_2 would remain in Ex. 5.3 in the text. The total mass of the CO and H_2 reactants was (356 g + 65.0 g) = 421 g and the amount of CH_3OH calculated to be formed based on the limiting amount of CO available was 407 g, leaving a difference of 14 g for g H_2 remaining. However, the authors chose to illustrate the steps of the more general approach outlined above, the approach that can be used with reactions more complicated than x A + y B → z C . You also can refer to the in-text consideration of $SiCl_4(\ell)$ + 2 Mg(s) → Si(s) + 2 $MgCl_2$(s) to see a second example of how to follow the steps outlined above.

Finally, *you might find it easier to conduct a limiting reactant calculation of maximum amount of product that can be formed by simply doing g reactant → mol reactant → mol product → g product calculations starting with each reactant and choosing the smallest value of g of product as the maximum amount that can be produced. Why is the smallest value the maximum amount? It is based on the limiting reactant, and no more product can be formed after the limiting reactant is completely consumed.* This method of solving a limiting reactant problem is illustrated in PROBLEM SOLVING TIPS AND IDEAS 5.4 in the text. *However, the procedure for calculating the mass of excess reactant remaining still would be as outlined in the preceding paragraph.*

5.3   Percent Yield

The amount of product that is calculated to be formed in a chemical reaction represents the maximum quantity of product that can be formed and is called the **theoretical yield**. However, there are numerous reasons why this amount might not be obtained in the laboratory. The quantity of product that is actually obtained is called the **actual yield**. The **percent yield** can be calculated using

$$\text{Percent yield} = \frac{\text{Actual yield}}{\text{Theoretical yield}} \times 100\% \qquad (5c)$$

If the percent yield is known for a process and there is a need to obtain a certain given number of grams of product, the percent yield formula can be solved for the theoretical yield that would give that actual amount and that theoretical yield can, in turn, be used to calculate the necessary grams of reactants to use.

5.4   Chemical Reactions and Chemical Analysis

One important use of chemical analysis involves analyzing the combustion products of carbon-hydrogen compounds to be able to determine empirical formulas. The combustion converts the carbon to carbon dioxide in the presence of sufficient oxygen, and the hydrogen is converted to water, so it is necessary to determine the amount of each formed. *If the compound contains only carbon and hydrogen, the masses of CO_2 and H_2O can be converted directly to moles of C and moles of H, and the C to H mole ratio can be calculated to obtain the empirical formula of the compound following the procedure that is used in Ex. 5.5 in the text. If the compound also contains oxygen or nitrogen or any other element, the moles of C and moles of H also need to be converted to grams of C and grams of H, so these quantities can be subtracted from the total mass of the sample to obtain the mass and moles of that third element, as in Ex. 5.6 in the text.*

An empirical formula can be converted to a molecular formula when the molar mass is known from other experiments. *A molecular formula is always a whole-number multiple of an empirical formula, and the value of the whole number is given by the ratio (molar mass/empirical formula molar mass).* If the simplest formula is CH_2 and the experimental molar mass is 85.1 g/mol, the multiple is (85.1 g/mol)/(14.0 g/mol) $\approx$ 6 and the molecular formula is C_6H_{12}.

5.5   Working With Solutions

Reactions are often carried out in solution because reactions between reactants in solution are often faster than reactions between solid reactants. In addition, many reactions naturally occur in solution, as in our bodies and in our environment. In these cases, it is desirable to use a concentration unit that relates moles of substance to volume of solution rather than mass.

Solution Concentration: Molarity

The substance that is dissolved in forming a solution is called the **solute**, and the dissolving medium is called the **solvent**. The amount of solute that is dissolved in a given volume of solvent is given by the **concentration** of the solution. Solution concentration is usually given as moles of solute per liter of solution, and this is called the **molarity** of the solution.

$$\text{Molarity, M} = \frac{\text{moles of solute}}{\text{liter of solution}} \qquad (5d)$$

The terms *moles per liter* and *molar* are used interchangeably, and brackets are commonly placed about the substance for which the molar concentration is being given, as with $[KMnO_4]$ = 1.0 M and $[SO_4^{2-}]$ = 0.36 M.

Preparing Solutions of Known Molarity:

Starting With A Solid Solute:

To prepare a solution of desired molarity and volume starting with a solid solute, the required grams of solute are weighed and dissolved in a volume of solvent that is smaller than the desired final volume. The resulting solution is then diluted to the desired final volume by adding more solvent. The required number of grams of solute should never be added directly to the desired final volume of solvent, because volume generally changes when solutions are made, and molarity refers to moles of solute per liter of *solution,* not moles of solute per liter of

---

solvent.

The required grams of solute can be calculated using:

L_{soln} · M · *M* solute

because the definition of molarity tells us multiplying M **by** L_{soln} **gives moles of solute, and moles can be converted to grams by multiplying by molar mass.** In other words,

L_{soln} · $\dfrac{\text{moles of solute}}{\text{liter of solution}}$ · $\dfrac{\text{grams of solute}}{\text{mole of solute}}$ = grams solute

Starting With A Concentrated Solution:

Solutions of desired molarity and volume can also be prepared by starting with concentrated solutions and adding solvent until the desired molarity and volume are reached. This method of preparation is called the **dilution method.**

The number of moles of solute remain the same as solvent is added during dilution. Thus,

moles of solute in   =   moles of solute
concentrated solution   in dilute solution

$$M_c \cdot L_c = M_d \cdot L_d \tag{5e}$$

can be used to calculate the volume of concentrated solution of known molarity that is required to prepare a given volume of solution of desired lower molarity. In addition, this relation can be used to calculate any given variable, provided the values of the other three variables are known. For example, it can also be used to calculate the molarity of the dilute solution, M_d, from M_c, L_c, and L_d, as illustrated in Ex. 5.7 in the text.

The concept of equivalent moles of solute before and after dilution was used to derive Expression 5e in terms of molarity and liters of solution. However, **mL can be used in place of L for both the dilute and concentrated solutions**, because the conversion factor for converting from mL to L would be the same on both sides of the equation and thus cancel out.

$$M_c \cdot mL_c \cdot \frac{1 \text{ L}}{10^3 \text{ mL}} = M_d \cdot mL_d \cdot \frac{1 \text{ L}}{10^3 \text{ mL}}$$

Indeed, **a more general and more useful form of the dilution equation is:**

$$C_1 \cdot V_1 = C_2 \cdot V_2 \qquad (5f)$$

where C_1 and C_2 represent any consistent set of volume dependent concentrations (molarities, parts per million (ppm), etc.) and V_1 and V_2 represent any consistent set of volumes (liters, milliliters, gallons, etc.).

5.6   Stoichiometry Of Reactions In Aqueous Solution

The definition of molarity tells us that the number of moles of solute in any given volume of solution of known molarity is given by:

$$\text{moles of solute} = L_{soln} \cdot M \qquad (5g)$$

This gives us another entry into the logic pattern for conducting stoichiometric calculations for chemical reactions that was introduced in 5.1, namely:

$$g\ A \longrightarrow mol\ A \longrightarrow mol\ B \longrightarrow g\ B \qquad (5h)$$
$$\uparrow$$
$$L_A\ \&\ M_A$$

This pattern can be used, for example, to calculate the number of grams of any solid that will react with a given volume of a solution of a solute of known molarity. Notice how the steps that were given in the text for calculating g $CaCO_3$ that would react completely with 25 mL of 0.750 M HCl follow this pattern by using $L_{HCl} \cdot M_{HCl}$ to calculate moles HCl, using the stoichiometric factor 1 mol $CaCO_3$/2 mol HCl to convert mol HCl to mol $CaCO_3$ and then using mol $CaCO_3 \cdot M$ $CaCO_3$ to convert mol $CaCO_3$ to g $CaCO_3$.

Titrations:

The determination of the quantity of a given constituent in a mixture is the goal of procedures used for **quantitative analysis**. A procedure that is commonly used in the laboratory portion of a general chemistry course is that of **titration**. A solution of known concentration is placed in a **buret**, which is a volumetric measuring device that is calibrated in divisions of 0.1 mL and is read by interpolating to the nearest 0.01 mL. A solution of a known quantity of the substance to be analyzed is placed in a flask, along with an **indicator**, which is a substance that changes color as close as possible to the point at which a stoichiometrically equivalent amount of reactant has been added from the buret. The point at which a stoichiometrically equivalent amount of reactant has been added is called the **equivalence**

point of the reaction, and the point at which the indicator changes color is called the **end point** of the titration, because that is the point at which we end the titration. *Indicators are selected on the basis of ability to change color as close to the equivalence point as possible.*

> Note: There also are instances in which the solution of unknown concentration is placed in the buret, and a known quantity of a solid of known composition or a known quantity of a solution of known concentration is placed in the flask. See Ex. 5.10 in the text.

The process of titration can be used to determine the percentage of a compound in an impure solid sample. The percentage of $CaCO_3$, for example, in impure limestone could be determined by titrating a weighed sample with a hydrochloric acid solution of known concentration. (*When we say "by titrating with" we mean the reactant we are "titrating with" is placed in the buret.*) The procedure for calculating the grams of $CaCO_3$ in the impure sample would be the same as that described above; except, in this case, the volume of HCl would be the volume of HCl actually delivered from the buret. The percentage of $CaCO_3$ in the ore would be calculated using (g $CaCO_3$/g sample) · 100%. Ex. 5.9 in the text uses an example in which a sample containing a solid acid is analyzed by titration with a *standard* solution of a strong base.

The procedure by which the concentration of an analytical reagent is determined is called **standardization**, and there are two general approaches. One approach involves weighing accurately a sample of a dry, pure solid acid or base, known as a **primary standard**, and then titrating this sample with a solution of the base or acid to be standardized (See Ex. 5.10 in the text.). *The other approach to standardizing a solution is to titrate it with another solution that has already been standardized. The goal of each of these standardization procedures is to determine the molar concentration of solute in the solution being standardized, the ratio of moles of solute to liters of solution. The logic pattern given above can be extended to include both M_B and L_B:*

$$\text{g A} \longrightarrow \text{mol A} \longrightarrow \text{mol B} \longrightarrow \text{g B} \qquad (5i)$$

$$L_A \ \& \ M_A \qquad M_B \qquad L_B$$

Then, since moles of solute = L · M, M_B = mol B/L_B and L_B = mol

B/M_B and either M_B or L_B can be calculated readily. **The mathematical operations for calculating M_B using conversion factors are:**

$$L_A \cdot M_A \cdot \frac{y \text{ mol B}}{x \text{ mol A}} \cdot \frac{1}{L_B} \qquad (5j)$$

and those for calculating L_B are:

$$L_A \cdot M_A \cdot \frac{y \text{ mol B}}{x \text{ mol A}} \cdot \frac{1}{M_B} \qquad (5k)$$

The latter expression can be used to calculate the volume of solution of known molarity of reactant B that contains the amount of B that is needed to completely react with the A in a given volume of a solution of known molarity of A.

Oxidation-reduction reactions can also be used for chemical analysis because many of these reactions go rapidly to completion in aqueous solution, and methods exist for finding the equivalence points for oxidation-reduction reactions, as well as acid-base reactions. Ex. 5.11 in the text involves an oxidation-reduction reaction in which the addition of the first excess of a colored reactant (purple $KMnO_4$) from the buret serves as the indicator.

5.7   Key Expressions

Logic pattern for conducting stoichiometric calculations for chemical reactions:

$$g \; A \longrightarrow mol \; A \longrightarrow mol \; B \longrightarrow g \; B$$

$$L_A \; \& \; M_A \qquad M_B \qquad L_B$$

Percent yield expression:

$$\text{Percent yield} = \frac{\text{Actual yield}}{\text{Theoretical yield}} \times 100\%$$

Molarity expression:

$$\text{Molarity, M} = \frac{\text{moles of solute}}{\text{liter of solution}}$$

Dilution expression:

$$C_1 \cdot V_1 = C_2 \cdot V_2$$

✳ Try preparing "flash cards" for each of these key expressions and memorizing them. Also try conducting the mathematical operations for conversions occurring with stoichiometric calculations for chemical reactions until they become so familiar that you can do them without taking too long to think about them. In the meantime, refer to Expressions 5b, j and k.

5.8 Hints For Answering Questions and Solving Problems

One of the most important keys for solving problems is being able to identify the type of problem that is being presented. If the problem:

* *gives the grams or moles of one substance participating in a chemical reaction and asks for the grams or moles of another substance participating in the same reaction, you should*

    * make sure the equation for the reaction is balanced and then

    * refer to the logic pattern that was given as Expression 5a on p. 61 and the mathematical operations for carrying out the necessary conversions that were outlined on p. 61.

* *gives the grams or moles of two or more reactants and asks for the maximum number of grams or moles of product that can be formed, you should realize that the problem is a limiting reactant problem and*

    * make sure the equation for the reaction is balanced,

    * follow the procedures outlined in 5.2 for determining which reactant is the limiting reactant and base your calculation of the maximum amount of product that can be formed on the amount of the limiting reactant, and

    * follow the procedures outlined in 5.2 for determining the amount of excess reactant, if that information is requested.

> Note: The alternative method of using
> *g reactant-to-g product* conversions
> with both reactants and selecting the
> smallest calculated grams of product
> as the maximum amount of product is
> <u>highly recommended</u>. See the last
> paragraph of 5.2 and also PROBLEM
> SOLVING TIPS AND IDEAS 5.4 in the
> text.

• *gives the grams of one or more of the reactants and the grams of one of the products obtained in a reaction and asks you to calculate the percent yield for the reaction, you should*

    • make sure the equation for the reaction is balanced,

    • proceed to calculate the theoretical yield of the product using either of the methods outlined above, depending on whether it is a limiting reactant case or not, and

    • use Expression 5c to calculate the percent yield.

• *gives the grams of a carbon-hydrogen containing compound and the grams of CO_2 and H_2O formed by its combustion and asks for the simplest formula and perhaps the molecular formula of the compound, you should follow the procedure outlined in 5.4 and illustrated in either Ex. 5.5 or 5.6 in the text.*

• *gives the grams of solute and volume of solution and asks for molarity <u>or</u> gives the volume of solution and molarity of the solution and asks for grams or moles of solute, you should realize the problem involves the definition of molarity and refer to Expression 5d and the discussion in the subsection immediately following Expression 5d.*

• *gives either two volumes and one concentration <u>or</u> one volume and two concentrations for solutions of the same substance and asks for either a concentration or a volume, respectively, for that substance, you should realize the problem is a dilution problem based on Expression 5f.*

- *gives the grams of one substance and asks for the volume or molarity of a solution of another substance or gives the volume and molarity of a solution of one substance and asks for the grams, volume of solution or molarity of solution of another substance, you should realize the problem involves solution stoichiometry and*

  - make sure the equation for the reaction is balanced,

  - refer to the logic pattern for reactions in solution.

> Note: When the given information falls into this last category and the statement of the problem asks for the percentage by mass of a component in a sample, you should realize that you will need to calculate grams to be able to use (g component/g sample) x 100% to calculate the desired percent.

* Try preparing "flash cards" listing the characteristics of each of these seven types of problems, and memorizing them to assist you in identifying the type of problem that is being presented. Also try preparing "flash cards" listing steps to be used in solving these types of problems and memorizing them and using them until they become so familiar that you can do them without taking too long to think about them.

CHAPTER 6

ENERGY AND CHEMICAL REACTIONS

Some chemical reactions occur naturally whereas others only occur under the influence of a constant external source of energy. Some occur to only a small extent whereas others go to completion. Furthermore, some occur slowly whereas others occur explosively fast. **Chemists, therefore, are concerned with: can a reaction occur, and if it can, how far will it go and how fast will it go.**

Chemical reactions can be classified as being product-favored or reactant-favored, depending on how far they go. If a reaction is a **product-favored reaction**, most of the reactants will be converted to products naturally, although it may take a very, very long time. The conversion of diamonds to graphite, for example, is *product-favored*, but the process is so very slow that diamonds are still being bought as jewelry. **Product-favored reactions usually evolve energy. The following types of reactions that were described in Chapter 4 are examples of product-favored reactions:**

- Reactions that form a precipitate,

- Reactions between acids and bases that form the weak electrolyte water,

- Reactions that produce a gas, such as the reaction of a metal carbonate with an acid, and

- Oxidation-reduction reactions involving the combination of metals with oxygen or the halogens to form metal oxides and halides, respectively, and those involving the combustion of organic carbon-containing compounds.

On the other hand, reactions having very little tendency to occur without the continuous input of energy from an external source are **reactant-favored reactions.** Fortunately, the reaction $N_2(g) + O_2(g) \rightarrow 2\ NO(g)$ is one of these reactions, and this is why nitrogen and oxygen, the two main components of the atmosphere, do not react with each other appreciably under ordinary conditions to form nitrogen oxides, which contribute to the formation of photochemical smog and acid rain. This reaction does, however, take place under the high temperature conditions found in automobile engines and power plants and under the influence of bolts of lightning. Thus, **the input of energy can**

cause the reactants of reactant-favored reactions to be converted to appreciable quantities of products.

Energy is clearly a factor in determining whether reactions are product- or reactant-favored. The transfer of energy as heat or thermal energy is a major theme of **thermodynamics**, the study of heat and work. The goal of Chapter 6 is to introduce some basic concepts about thermal energy and some basic thermodynamic terminology, to teach you how to calculate heat flows accompany both changes of temperature and changes of physical state, to introduce the First Law of Thermodynamics giving the relation between change in internal energy, heat and work, to teach you how to use Hess's Law and enthalpies of formation to calculate enthalpy changes for physical and chemical changes, to teach you how to use bomb calorimetry and coffee-cup calorimetry data to calculate heat flows under conditions of constant volume and constant pressure, respectively, and thus changes in internal energy and enthalpy, respectively, and present some information about energy resources and consumption.

6.1   Energy: Its Forms and Its Units

Energy is defined as the capacity to do work or supply heat. *Energy can be assigned to one of two classes: kinetic or potential.* An object has **kinetic energy** because it is moving. Examples include: the *thermal energy* associated with the motion of atoms, molecules or ions, the *mechanical energy* associated with the motion of macroscopic objects, the *electric energy* associated with electrons moving through conducting wires and *sound*, because it involves compression and expansion of the spaces between molecules. **Potential energy** is energy that an object has due to its position or condition. Examples include: *chemical potential energy* resulting from attractions between atomic nuclei and electrons in molecules and positive and negative ions, cations and anions, in ionic compounds and *gravitational energy* due to position above some reference point. *Kinetic energy can be converted to potential energy, and potential energy can be converted to kinetic energy. The kinetic energy of an object in motion is given by: $KE = \frac{1}{2}mv^2$, where m represents the mass of the object and v represents its velocity.*
✱ Try preparing three separate "flash cards" summarizing the main classes of energy and their characteristics, the examples of each main class of energy and the formula for calculating KE and memorizing them.

Temperature, Heat, and the Conservation of Energy

Heat is associated with temperature, but heat is not the same as temperature. The thermal energies of the particles of an object increase when an object is heated, and this is reflected in the temperature increase of the object. **Energy has been added to the object as heat, but heat is not a substance that is contained in the object.** Hence, we should never say that an object that undergoes a temperature increase by heating contains more heat.

Energy transfer, as heat, happens whenever two objects at different temperatures are brought into contact. Energy always transfers spontaneously from the hotter to the cooler object until the two objects reach the same temperature.

There are several important aspects of thermal energy that must be understood:

- The more energy an atom or molecule has, the faster it moves, but the atoms or molecules in any given sample of a substance can have differing energies and, therefore, have differing speeds.

- The average speed of the atoms or molecules in a substance is related to the temperature of the substance.

- The total thermal energy in a substance is the sum of the individual energies of all the atoms or molecules in that substance.

This means the total thermal energy of a substance depends on the temperature and the amount of the substance. Thus, it is possible for a large container of water to contain more thermal energy than a cup of water that is at a much higher temperature. ✳ Try preparing a "flash card" listing these important aspects of thermal energy and memorizing them.

Energy Units

The thermal unit of energy is the calorie. A **calorie** was originally defined as the amount of energy, transferred as heat, that was required to raise the temperature of exactly 1 g of pure liquid water by exactly 1 degree Celsius, actually from 14.5°C to 15.5°C. The thermal energy unit that is often used in science is the kilocalorie, abbreviated kcal. One kilocalorie is equivalent to 1000 cal and one nutritional calorie, 1 C. That is, 1 kilocalorie = 1 kcal = 1000 cal = 1 C.

The mechanical energy unit of energy is the joule. A **joule** is the amount of kinetic energy that is possessed by a 2.0 kg object moving at a velocity of 1.0 m/s, because KE = $\frac{1}{2}mv^2$ = $\frac{1}{2}(2.0$ kg$)(1.0$ m/s$)^2$ = 1.0 kg·m^2/s^2 = 1.0 J. **The unit of energy that is widely used in science is the kilojoule, abbreviated kJ. One**

calorie is now defined as exactly 4.184 J, so one kilocalorie is equivalent to 4.184 kJ. ✷ Try preparing a "flash card" summarizing the bases for the thermal and mechanical units of energy and memorizing them.

6.2 Heat Capacity and Thermal Energy Transfer

The **specific heat capacity, c,** of a substance, which is often just called its **specific heat**, is defined as the amount of energy that must be transferred as heat to raise the temperature of exactly 1 g of that substance by exactly 1°C (or equivalently 1 kelvin). *Mathematically then,*

$$c = \frac{\text{quantity of heat supplied}}{\text{(mass of substance)} \cdot \text{(temperature change)}} \qquad (6a, 6.1)$$

$$= \frac{q \ (J)}{[m \ (g)] \cdot [\Delta T \ (K)]} \qquad (6b)$$

where q is the commonly used symbol for heat and the capital Greek letter delta, Δ, means "change in", so the units for specific heats are joules per grams times kelvins, J/g·K or J·g^{-1}·K^{-1}. One method for determining the specific heat of a metal is described in Ex. 6.3 in the text.

The higher the specific heat of a substance, the greater the amount of energy that must be transferred as heat to raise or lower the temperature of 1 g of the substance by 1 kelvin. Water has one of the highest specific heat values known, 4.184 J/g·K. Thus considerable heat must be transferred to raise or lower the temperature of large bodies of water, so they exhibit a moderating influence on the temperatures of air and objects around them.

The equation defining specific heat can be solved for q giving:

$$q = \text{(specific heat)(mass)(temperature change)} \qquad (6c, 6.2)$$

$$= \text{(c in J/g·K)(mass in grams)}(\Delta T \text{ in kelvins}) \qquad (6d)$$

Changes represented by Δ in chemistry always refer to (final value - initial value), so ΔT = (final temperature - initial temperature) = $(T_f - T_i)$. When energy is transferred into a substance as heat resulting in a temperature increase, $T_f > T_i$ and both ΔT and q have a positive sign. Conversely, when energy is transferred out of a substance as heat, $T_f < T_i$ and both ΔT and q have a negative sign. These important relations are summarized in A Closer Look: Sign Conventions in Energy Calculations in the text. ✷ Try preparing a "flash card" showing this information and memorizing it. *Also note that the implications of these sign conventions for equating thermal energy changes for thermal energy transfers between substances. See Ex. 6.3 and Problem Solving Tips and Ideas 6.3 in the text.*

6.3 Energy and Changes of State

Energy transfers also occur when matter is transformed from one form to another in the course of physical and chemical changes. *When a solid melts and a liquid boils, energy is required to enable the particles (atoms, molecules or ions) to overcome the attractions of their neighboring particles and move farther apart. This involves changes in potential energy, so changes of state always take place at constant temperature, as shown in Fig. 6.9 in the text.* ∗ Try preparing a "flash card" summarizing this important information and memorizing it.

The quantity of heat required to melt ice, called the **heat of fusion,** is 333 J/g at 0°C. The quantity of heat required to vaporize liquid water, called the **heat of vaporization**, is 2260 J/g at 100°C. *The heat of vaporization is usually much greater than the heat of fusion for any given substance, because the change from liquid to vapor involves a much greater change in distances between particles than the change from solid to liquid.* On the other hand, *heat is liberated when these processes are reversed; that is, when steam condenses to liquid water and when liquid water freezes.* This is why steam burns often are more severe than hot water burns and why citrus growers spray their trees with water when the temperature is likely to go below the freezing point.

Heat is required for melting and vaporizing substances, and these processes are said to be **endothermic**. On the other hand, heat is liberated with condensing and freezing, and these processes are said to be **exothermic**. *Thus, the sign of q for heat transfers involving melting and vaporizing is positive and that for heat transfers involving condensing and freezing is negative. In the case of reversible changes involving the same amount of substance, the quantity of heat involved is numerically the same; only the sign of q changes.* The q associated with the condensing of water, therefore, is -2260 J/g at 100°C and that associated with the freezing of water -333 J/g at 0°C. ∗ Try preparing a "flash card" giving the names of all changes of state and the signs of q for those changes of state and memorizing it.

The actual quantity of heat involved in any change of state, q, varies with the grams of substance and is given by the product of mass in grams times the appropriate heat flow in J/g. See Ex. 6.4 in the text.

$$q = \text{(mass in grams)(change of state heat flow in J/g)} \qquad (6e)$$

Note: *The sign of q is positive for every endothermic change and negative for every exothermic change.*

6.4 Enthalpy

It is important to distinguish between the system and the surroundings when discussing heat flow. In chemistry, the **system** only includes the substances involved in the physical or chemical changes of interest, the reactants and the products. The reaction container and the rest of the universe comprise the **surroundings**. Introducing this terminology allows us be more specific about the nature of *endothermic* and *exothermic* changes. *The prefix "endo" means "in" and the prefix "exo" means "out".* Thus, **endothermic changes** involve the transfer of energy as heat from the *surroundings* to the *system* whereas **exothermic changes** involve the transfer of energy as heat from the *system* to the *surroundings*. The relations between direction of heat transfer,

sign of q, and type of change are summarized in the text. * Try preparing a "flash card" containing this information and memorizing it. Also try drawing the illustration of these concepts given in the text to reinforce your understanding of this information.

The **internal energy, E,** in any given chemical system can be viewed as the sum of the potential and kinetic energies associated with the system. *When energy is transferred into or out of a system as heat under constant volume conditions, the increase in internal energy of the system, ΔE, is given by q: $\Delta E = q_v$, where the subscript v indicates the transfer has occurred at constant volume. However, when the system is allowed to expand and do work on its surroundings or contract as the surroundings do work on the system, energy is transferred out of or into the system, respectively, by work, symbolized by w. Thus, under these conditions work also must be taken into account. The change in internal energy of the system is therefore given by:*

$$\Delta E = q + w \qquad (6f, 6.3)$$

which states the internal energy of the system is changed by the quantity of energy transferred as heat and by the quantity of

---

energy transferred as work.

The signs for q and w can thus be rationalized on the basis of the predicted impact on the internal energy of the system.

Energy Transferred As:	Impact On Internal Energy Of System	Signs Of q and w
Heat from surroundings to system	increase	q is +
Work from surrounding to system by surroundings doing work on system	increase	w is +
Heat from system to surroundings	decrease	q is -
Work from system to surroundings by system doing work on surroundings	decrease	w is -

Energy transfers to the system increase the internal energy of the system and require the use of positive signs for both q and w with Expression 6.4 whereas energy transfers out of the system decrease the internal energy of the system and require the use of negative signs for both q and w with Expression 6.4. ✶ Try preparing a "flash card" summarizing this information and memorizing it, but also be able to give the rationale for the signs of q and w based on the predicted impact on the internal energy of the system.

The expression $\Delta E = q + w$ is a mathematical statement of the **First Law of Thermodynamics**, also called the Law of Conservation of Energy. In other words, **the total amount of energy in the universe is constant. Energy may be transferred as work or heat, but no energy can be lost, nor can heat or work be obtained from nothing. All the energy transferred between a system and its surroundings normally must be accounted for as heat and work;** the text cites the release of radiant energy by a glow stick as a "glowing" exception to total transfer of energy as heat and work.

The heat transferred into or out of a system at constant pressure, q_p, equals a quantity called the **enthalpy change**, symbolized by **ΔH**. **Thus, we have:**

$\Delta E = q_v = $ *heat transferred into or out of a system at constant volume*

and

$\Delta H = q_p = $ *heat transferred into or out of a system at constant pressure*

At constant pressure $\Delta E = q + w$ becomes $\Delta E = q_p + w = \Delta H + w$, showing that ΔH accounts for all of the energy transferred at constant pressure except the amount that is used for expansion of the system or the amount that is gained by the forced contraction of the system. It turns out that the amount of energy associated with work of expansion and contraction is only significant when there a large change in the number of moles of gas during chemical changes and physical changes of state. This means, $\Delta E = \Delta H$ when there is no change in the number of moles gases in the process of interest, and that $\Delta E \approx \Delta H$ with most other chemical changes and physical changes of state.*

The enthalpy change, ΔH, for any process is given by:

ΔH = (enthalpy of the system at the end of the process)
    − (enthalpy of the system at the start of the process)

$= H_{final} - H_{initial}$ (6g, 6.4)

This means that when the enthalpy of the final state of the system is greater than that of the initial state of the system, ΔH and q_p are positive. There has been an energy transfer as heat from the surroundings to the system, and the process is endothermic with respect to the system. Conversely, when the enthalpy of the final state of the system is lower than that of the initial state of the system, ΔH and q_p are negative and the process is exothermic with respect to the system. We emphasize with respect to the system, because the ΔH values for the system and the surroundings are numerically the same, but their signs are opposite. Suppose you give a friend a five dollar bill. The change would be -\$5 for you and +\$5 for your friend. The system and surroundings transfer heat in an analogous manner. *In summary,*

Energy Transferred as Heat at Constant Pressure From:	Sign of q_p	Sign of ΔH	Nature of Process With Respect to System
Surroundings <u>to</u> system	+	+	endothermic
System <u>to</u> surroundings	−	−	exothermic

The signs of q_p and ΔH must necessarily be the same, because q_p = ΔH. The signs of q_p also must necessarily agree with those given in the summary above, because $\Delta E = q_p + w$ at constant pressure, and the impact of the energy transfer as heat on the internal energy of the system, E, is predicted to be the same.
✻ Try preparing a "flash card" summarizing this important information and memorize it.

**There are several additional key ideas pertaining to enthalpy
changes that can be shown with physical changes of state:**

• For changes that are the reverse of each other, the ΔH
values are numerically the same, but their signs are
opposite. We supply heat to melt ice, and the same amount
of heat is released as the same amount of water freezes.

• The change in energy or enthalpy is directly proportional
to the quantity of material undergoing a change. We need to
supply more heat to melt more ice. **The quantity of heat
involved in any change of state is given by the product of
mass in grams times the appropriate heat flow in J/g
(Expression 6.3) or the product of moles of substance times
the appropriate enthalpy change, ΔH, in kJ/mol (Ex. 6.5 in
the text.)**

• The value of ΔH is always associated with a balanced
equation for which the coefficients are read as moles, so
the equation shows the macroscopic amounts of material to
which ΔH applies. This means balanced equations having
fractional coefficients are acceptable in thermochemistry
because one can have fractions of moles; fractions of
molecules do not exist, so we always use equations with
smallest whole-number coefficients with stoichiometry.

✱ Try preparing a "flash card" summarizing these key ideas and memorizing them.

6.5   Enthalpy Changes For Chemical Reactions

**The products of a chemical reaction represent the final state of
the system, and the reactants represent the initial state of the
system. Thus, for chemical reactions, $\Delta H = H_{products} - H_{reactants}$:
The key ideas pertaining to enthalpy changes for physical
changes of state are applicable to chemical reactions, as well.**
You will commonly hear the term "heat of reaction" used interchangeably with the term "enthalpy of reaction" but should
remember that only the heat of reaction at constant pressure, q_p,
is equivalent to the enthalpy change.

The direction of heat transfer provides an important clue as to
whether a chemical reaction will be reactant-favored or product-
favored. **At room temperature, most exothermic reactions are
product-favored.**

6.6   Hess's Law

Energy conservation is the basis for **Hess's Law**, which states that, *if a reaction is the sum of two or more other reactions, then ΔH for the overall process must be the sum of the ΔH values of the constituent reactions.* The application of Hess's Law enables us to calculate enthalpy changes for reactions or changes of state that cannot be conducted conveniently in the laboratory. **Enthalpy diagrams** can be used to give visual representation of concepts involved with the application of Hess's Law.

6.7   State Functions

Hess's Law works because the enthalpy change for a reaction is a **state function**, a quantity whose value is determined only by the present state of the system. *The enthalpy change for a chemical or physical change does not depend on the path that is taken to go from the initial state to the final state. It is only based on the difference in enthalpy between the final and initial states.* We usually do not even know the enthalpies of the individual participants in a physical or chemical change; we only measure changes in enthalpies for physical and chemical changes.

6.8   Standard Enthalpies of Formation

ΔH values depend on pressure and temperature, so it is necessary to specify the values of both of these quantities when a table of data is made. Usually a pressure of 1 bar (1 bar = 0.98692 x standard atmospheric pressure measured at sea level) and a temperature of 25°C are specified, although other temperature conditions are used.

When a reaction occurs with all the reactants and products under standard conditions, the observed, or calculated, enthalpy change is known as the **standard enthalpy change of reaction,** ΔH°_{rxn} where the superscript $^{\circ}$ indicates standard conditions. Furthermore, the **standard state** of an element or a compound is defined as the most stable form of the substance in the physical state in which it exists at 1 bar and the specified temperature. Using standard conditions and standard states as reference points enables us to conveniently and accurately compare enthalpy changes for various changes.

The standard enthalpy change for a reaction in which one mole of a compound is formed directly from its elements is called the **standard molar enthalpy of formation, ΔH_f^O,** of the compound. The subscript f indicates that one mole of the compound has been formed in its standard state from its elements in their standard states. *ΔH_f^O values are listed in Table 6.2 and Appendix K in the text. Notice that the ΔH_f^O values for most compounds are negative indicating that reactions for forming these compounds from their elements under standard conditions is generally product-favored. On the other hand, the standard enthalpies of formation for elements in their standard states are zero, because forming an element in its standard state from the same element in its standard state involves no chemical or physical change.*

The standard enthalpy change for any reaction can be calculated when the standard enthalpies of formation of all the reactants and products are known. The equation for doing this is:

$$\Delta H_{rxn}^O = \Sigma \, [\Delta H_f^O(\text{products})] \ - \ \Sigma \, [\Delta H_f^O(\text{reactants})] \qquad (6h, 6.6)$$

where the capital Greek letter sigma, Σ, means to "take the sum." Applying Expression 6h to the general equation aA + bB $\rightarrow$ cC + dD gives:

$$\Delta H_{rxn}^O = c\Delta H_f^O[C] \ + \ d\Delta H_f^O[D] \ - \ a\Delta H_f^O[A] \ - \ b\Delta H_f^O[B] \qquad (6i)$$

Notice how it is necessary to multiply the ΔH_f^O value for each substance by the numbers of moles of that substance in the balanced equation, because ΔH_f^O values correspond to forming one mole and are, therefore, tabulated in terms of kJ/mol (See Table 6.2 and Appendix K in the text.). Once the value of ΔH_{rxn}^O is known for a reaction, the enthalpy change associated with the reaction of any quantity of reactant or the formation of any quantity of product can be calculated. See Ex. 6.8 in the text to learn how to write **thermochemical factors** relating enthalpy change to moles and using them to do these kinds of calculations.

6.9 Determining Enthalpies of Reaction

The heat evolved by a chemical reaction can be determined by a technique called **calorimetry**. When determining heats of combustion or the caloric value of foods, the measurement is often done using a *combustion calorimeter*, which is also known as a **bomb calorimeter** (See Fig. 6.17 in the text.). A weighed sample is placed in an enclosed cylinder called a "bomb," and the bomb is filled with oxygen and lowered into a water bath. The reaction mixture of sample and oxygen is ignited by using an electrical spark, and the temperature increase of the bomb and surrounding water is measured electronically. **Applying the Law of Conservation of Energy gives:**

Heat evolved by   = -(Heat absorbed by the water
  the reaction   and the bomb)

$$q_{rxn} = -(q_{water} + q_{bomb}) \tag{6j}$$

where the minus sign is included because the heat released by the reaction is numerically equal to the heat gained by the bomb and surrounding water, but of opposite sign. In other words, the sign of q_{rxn} will be negative because the combustion reaction is exothermic. **Mathematically,**

$$q_{rxn} = -[(c_{water})(g\ water)(\Delta T) + (C_{cal})(\Delta T)] \tag{6k}$$

with ΔT necessarily being the same for the water and bomb. Because the bomb is rigid, the heat transfer is measured at constant volume and is therefore equivalent to ΔE, which is equal to q_v. However, because $\Delta E = \Delta H + w$, ΔH values can be calculated from ΔE values obtained with bomb calorimetry, though this is not done in this text.

The type of calorimeter that is usually used in a general chemistry laboratory is a constant pressure **coffee-cup calorimeter** (See Fig. 6.18 in the text.). *The heat that is evolved in a coffee-cup calorimeter is q_p and is a measure of the enthalpy change for the reaction, because $q_p = \Delta H$.*

In the case of coffee-cup calorimetry, the heat that is evolved by the reaction is gained only by the contents of the cup, the solution containing the reaction mixture, because virtually no heat is gained by the nested styrofoam coffee cups. Thus,

$$q_{rxn} = -q_{soln} \tag{6l}$$

$$q_{rxn} = -[(c_{soln})(g\ soln)(\Delta T)] \tag{6m}$$

The value that is obtained for q_{rxn} can be converted to an enthalpy change per mole of any reactant or product, as in Ex. 6.10 in the text. The relation $\Delta E = \Delta H + w$ can be used to calculate to calculate ΔE values from ΔH values obtained with constant pressure calorimetry, though this is not done in this text.

6.10 Applications of Thermodynamics

The text names several sources of energy, starting with biomass; three ways in which industry can save energy; and three desirable characteristics of fuels. ✳ Try preparing "flash cards" for each of these and memorizing them. The text also has a discussion of the potential of having a "hydrogen economy." ✳ Try preparing a "flash card" listing the advantages and disadvantages of using hydrogen as a fuel and memorizing them.

6.11 Key Expressions

Specific heat capacity expression:

$$c = \frac{\text{quantity of heat supplied}}{(\text{mass of substance}) \cdot (\text{temperature change})}$$

$$= \frac{q \ (J)}{[m \ (g)] \cdot [\Delta T \ (K)]}$$

Heat flow for temperature change expression:

$$q = (\text{specific heat})(\text{mass})(\text{temperature change})$$

$$= (c \text{ in } J/g \cdot K)(\text{mass in grams})(\Delta T \text{ in kelvins})$$

Heat flow for change of state expression:

$$q = (\text{mass in grams})(\text{change of state heat flow in } J/g)$$

$$= (\text{moles})(\text{change of state enthalpy change in } kJ/mol)$$

First Law of Thermodynamics mathematical expression:

$$\Delta E = q + w$$

Internal energy and enthalpy change expressions:

$$\Delta E = q_v$$

$$\Delta H = q_p$$

$$\Delta H = H_{final} - H_{initial}$$

Enthalpy of reaction - Enthalpy of formation expression:

$$\Delta H_{rxn}^{o} = \Sigma [\Delta H_f^o(\text{products})] - \Sigma [\Delta H_f^o(\text{reactants})]$$

Bomb calorimetry expression:

$$q_{rxn} = -[(c_{water})(g\ water)(\Delta T) + (C_{cal})(\Delta T)]$$

Coffee-cup calorimetry expression:

$$q_{rxn} = -[(c_{soln})(g\ soln)(\Delta T)]$$

6.12 Hints For Answering Questions and Solving Problems

Writing Thermochemical Equations:

Energy is produced and leaves the system as heat during exothermic chemical reactions. Thus, **heat can be considered to be a product in an exothermic chemical reaction and can be incorporated into the balanced chemical equation for the process**. This means

$$C_3H_8(g) + 5\ O_2(g) \rightarrow 3\ CO_2(g) + 4\ H_2O(\ell)\quad \Delta H° = -2220\ kJ$$

and

$$C_3H_8(g) + 5\ O_2(g) \rightarrow 3\ CO_2(g) + 4\ H_2O(\ell) + 2220\ kJ$$

are equivalent ways of writing the thermochemical equation for the exothermic complete combustion of three moles of propane under standard conditions. Conversely, **heat can be considered to be a reactant in an endothermic chemical reaction.**

Using Hess's Law:

The key to finding the algebraic combination of equations to use when working with Hess's Law is to:

• leave some blank space for writing the combination of equations to be used,

• writing the desired equation below a line at the bottom of the blank space, so it becomes the "target" equation,

• identifying one substance that appears in the target equation and in only one of the equations to be used,

• using the equation containing the substance in a manner that will yield the desired number of units of that substance in the proper position in the target equation, and

• systematically adding the other equations in a manner that causes substances appearing in the first and each succeeding equation to cancel if they are not part of the target equation.

In the case used in Ex.6.9 in the text, the target equation would be the net equation: $C(s) + 2 H_2(g) \rightarrow CH_4(g)$, suggesting the given equation $C(s) + O_2(g) \rightarrow CO_2(g)$ be used as written to provide one $C(s)$ on the left side for the target equation. The next step would have been to reverse the equation involving $CO_2(g)$ to provide one $CO_2(g)$ on the left side to cancel the $CO_2(g)$ formed by the first equation, because $CO_2(g)$ is not part of the target equation. Finally, the last remaining equation would multiplied by two to provide $2 H_2(g)$ for the left side of the target equation and simultaneously cancel all other substances appearing in the given equations but not part of the target equation. ✳ Try doing this on your own without relying too heavily on the steps just outlined. Also try using this sequential strategy with the series of equations given in Numerical Questions 42 and 56 in the text.

Identifying and Understanding State Functions:

Thermodynamic quantities that are state functions are represented by capital letter symbols, and changes in state functions are represented by using Δ. Thus, internal energy, E, and enthalpy, H, are thermodynamic state functions having changes designated as ΔE and ΔH, respectively, whereas heat, q, and work, w, are not. Consider the following scenario to appreciate this. The same change in the internal energy of a battery can be conducted by using electricity from the battery run an electric motor or by (foolishly) laying a wrench across the terminals of the battery. In the first case, the change in internal energy would appear primarily as electrical work, and in the second case the same change in internal energy would appear as heat. Thus, q and w are path dependent whereas change in internal energy is not. This is why the symbolic form of the First Law of Thermodynamics is $\Delta E = q + w$ and not $E = \Delta q + \Delta w$.

Writing Standard Molar Enthalpy of Formation Equations:

An equation showing the formation of one mole of a compound in its standard state from its constituent elements in their standard states is called a **standard molar enthalpy of formation equation**. *The key to writing these equations is to look at the formula of the compound, then write the formulas of the constituent elements in their standard states and use multiplying coefficients with these formulas to provide the necessary* *numbers of atoms of each element for one mole of the desired compound, remembering that fractional coefficients can be used in thermodynamics.*

The following will help you identify the standard states of the elements, though you can always consult Appendix K to determine the form of the element having $\Delta H_r^O = 0.0$ kJ/mol:

• The standard state of metals other that mercury is the solid state, though some metals can exist in several different crystal forms with one of these being most stable at 1 bar and 25°C. Mercury is, of course, a liquid under these conditions.

• The standard states of some of the more commonly encountered nonmetals are:

$$H_2(g) \qquad C(s, graphite) \qquad N_2(g) \qquad O_2(g) \qquad F_2(g)$$
$$P_4(s) \qquad S_8(s) \qquad Cl_{2(g)}$$
$$Br_2(\ell)$$
$$I_2(s)$$

where we have intentionally neglected to specify the particular crystalline forms of $P_4(s)$ and $S_8(s)$ for our purposes here.

Using the *Enthalpy of reaction - Enthalpy of formation* expression,

$$\Delta H_{rxn}^O = \Sigma \, [\Delta H_f^O(products)] \, - \, \Sigma \, [\Delta H_f^O(reactants)]:$$

The key to using this expression is to remember:

• to multiply the appropriate standard molar enthalpies of formation by the stoichiometric coefficients giving

$$\Delta H_{rxn}^O = c\Delta H_f^O[C] + d\Delta H_f^O[D] - a\Delta H_f^O[A] - b\Delta H_f^O[B]$$

for $aA + bB \rightarrow cC + dD$.

• the standard molar enthalpy of formation of any element in its standard state is zero.

• that the expression can be solved for any given ΔH_f^O value when the ΔH_{rxn}^O and all other ΔH_f^O values are known.

CHAPTER 7

ATOMIC STRUCTURE

Our currently accepted theory about the arrangement of electrons in atoms and thus about the chemical behavior of atoms is based on interactions between matter and electromagnetic radiation, especially visible light. This chapter traces the historical development of the understanding of the wave-particle dual nature of light, the concept of quantized energy states for electrons in hydrogen atoms and the wave-particle dual nature of electrons. It concludes with a consideration of the wave mechanical or quantum mechanical view of the atom and the arrangement of electrons in shells, subshells and orbitals.

7.1 Electromagnetic Radiation

In 1864, James Clerk Maxwell developed the currently accepted theory that *all forms of radiation are propagated through space as vibrating or oscillating electric and magnetic fields travelling at right angles to one another.* Hence, radiation, such as γ rays, x-rays, ultraviolet radiation, visible light, infrared radiation, microwaves and radio and television signals, is collectively called **electromagnetic radiation**. Our currently accepted theory about the arrangement of electrons in atoms and thus about the chemical behavior of atoms is based on interactions between matter and electromagnetic radiation, especially visible light. This is why it is important to learn some of the characteristics of waves.

The **wavelength** of a wave is the distance between any two corresponding parts of the wave, such as *crests* or *troughs*, and the symbol for wavelength is the Greek letter λ (lambda). The number of complete waves, complete cycles, passing through a point per second is called the **frequency** of the wave, and the symbol for frequency is the Greek letter ν (nu). The unit for frequency is usually written as s^{-1} (standing for 1 cycle/second or simply 1/second) and is now called the **hertz**, Hz. The maximum displacement of a wave from its axis of propagation is called the **amplitude** of the wave. Points of zero amplitude are called **nodes**; nodes always occur at intervals of $\lambda/2$ for standing waves, such as those illustrated in Fig. 7.2 in the text. ✱ Try preparing a "flash card" summarizing these characteristics of waves and their symbols and memorizing them.

The speed of a wave is given in terms of distance/second and is equal to the product of wavelength times frequency for the wave.

$$\frac{distance}{wave} \cdot \frac{waves}{time} = \frac{distance}{time}$$

In the case of electromagnetic radiation, the speed is the **speed of light** in a vacuum, symbolized by c, and that is 2.99792458 x 10^8 m/s (or 186,000 miles/s). *Hence,*

$$\lambda \cdot \nu = c = 2.998 \times 10^8 \text{ m/s} \qquad (7a, 7.1)$$

for all forms of electromagnetic radiation. Because the speed of light is given in m/s, *wavelengths must be converted to meters to use this expression.*

Visible light consists of a spectrum of colors ranging from red light at the long-wavelength end of the spectrum to violet light at the short-wavelength end of the spectrum. The full range of colors and electromagnetic radiation is given in Fig. 7.3 in the text. *You can remember the colors of visible light by using the phrase ROY G BIV, standing for red, orange, yellow, green, blue, indigo and violet.* Meanwhile, notice the relation between wavelength, frequency and energy that is given in Fig. 7.3 in the text: *The longer the wavelength, the lower the frequency* (in accord with Expression 7a) *and the lower the energy.* Thus, *the range of electromagnetic radiation is from the short wavelength, high frequency and high energy γ radiation to the long wavelength, low frequency and low energy radio and TV waves.* ✳ Try preparing a "flash card" summarizing the components of the electromagnetic spectrum and another "flash card" summarizing the relation between wavelength, frequency and energy and memorizing them.

The wave motion that has been discussed so far is that of *travelling waves*. However, the type of wave motion that is important to modern atomic theory is that of **standing,** or **stationary waves**. Standing waves are illustrated in Fig. 7.4 in the text. *Notice that*

- a standing wave is characterized by having two or more points of zero amplitude, two or more nodes. As with travelling waves, the distance between consecutive nodes is always $\lambda/2$.

- the standing wave ends must be nodes, so that $a = n(\lambda/2)$ where a is the distance from one end or *boundary* to the other and n is an integer $(1,2,3,...)$. This means only certain wavelengths are possible, and this is an example of **quantization** in nature.

7.2 Planck, Einstein, Energy and Photons

At the end of the 19th century, scientists assumed vibrating atoms in heated objects caused electromagnetic vibrations (light waves) to be emitted, and that these light waves could have any frequency along a continuously varying range. The classical wave theory predicted that as the object got hotter and acquired more energy, its color should shift to blue and then to violet and beyond, but no object was ever observed to do this, so the phenomenon became known as the "**ultraviolet catastrophe**."

In 1900, the German physicist, Max Planck, assumed there must be a minimum quantity of energy that could be emitted by an object at any given time and called this packet of energy a **quantum**. *Planck further stated the energy of a quantum is related to the frequency of the radiation by the equation*

$$E_{\text{quantum}} = h \cdot \nu_{\text{radiation}} \qquad (7b, 7.2)$$

where the proportionality constant h, called **Planck's constant**, has the value $6.6260755 \times 10^{-34}$ J·s.

The theory based on Planck's work is called the **quantum theory**. Using quantum theory, Planck was able to calculate the number of quanta of each frequency that are emitted by a heated object. The number of quanta emitted per second gives the intensity of the radiation, and because frequency is related to wavelength, Planck was able to calculate the spectrum of a heated object. *His results agreed very well with experimentally measured spectra and thus provided an acceptable explanation for the "ultraviolet catastrophe." The prediction of classical wave theory that continuous amounts of energy could be emitted was in error.*

Einstein and the Photoelectric Effect

Planck's theory was so radical that it was not well accepted at first, but after Albert Einstein used it to explain the photoelectric effect, Planck's quantum theory was firmly accepted. The **photoelectric effect** occurs when light strikes the surface of a metal and electrons are ejected. *Electrons are only ejected if light of some minimum frequency is used. If light of lower than the minimum frequency is used, no electrons are emitted no matter how intense (bright) the light. However, if the frequency of the incident light is greater the minimum value, increasing the light intensity causes more and more electrons to be ejected. Thus, Einstein assumed light could be described by both wave and particle-like properties* and that the *massless* particles of light, now called **photons**, carried the energy calculated by Planck's equation; that is, that the energy of each photon in a stream of photons is proportional to the frequency of its wave. *The more intense the beam of light, the more photons striking the surface of the metal and the greater the number of electrons ejected, provided the energy of the light exceeds the minimum energy.*

Energy and Chemistry: Using Planck's Equation

Planck's equation can be extended to express the energy of photons and therefore electromagnetic radiation in terms of wavelength. Using Expressions 7a and 7b together gives

$$E = h \cdot \nu = \frac{h \cdot c}{\lambda} \tag{7c}$$

This expression tells us that the energy of electromagnetic radiation increases as the wavelength decreases and the frequency increases. Thus, energy increases in the order infrared radiation, visible light and ultraviolet radiation.

7.3   Atomic Line Spectra and Niels Bohr

The final piece of information that played a major role in the development of the modern view of atomic structure was the observation that the light that is emitted by energetically excited atoms of an element gives a characteristic **line spectrum** when passed through a prism rather than a **continuous spectrum**. *The Rydberg equation*

$$1/\lambda = R(1/2^2 - 1/n^2) \quad n > 2 \tag{7d, 7.3}$$

was developed by Johann Balmer and Johannes Rydberg to calculate the wavelength of the three longest lines in the visible emission spectrum of hydrogen atoms. The equation was developed by empirically trying to fit the experimentally observed wavelengths to some (any) mathematical relation. This group of lines in the visible region of the spectrum is now called the **Balmer series** of lines.

The Danish physicist Niels Bohr was the first to propose a model for the hydrogen atom that could account for the line spectrum of hydrogen. Bohr assumed the electron of the hydrogen atom travelled in certain circular orbits about the nucleus corresponding to certain allowed energy levels, though classical physics predicted the atom could not exist this way. *Bohr's thinking led to the concept of "quantized" energy levels for electrons in atoms.* By combining his quantization postulate with the laws of motion from classical physics, *he was able to show* that the energy possessed by the single electron in the nth orbit of the H atom is given by

$$E_n = -\frac{Rhc}{n^2} \qquad (7e, 7.4)$$

wherw R is a proportionality constant, h is Planck's constant, and c is the speed of light. Each allowed orbit was assigned a unitless integer value of n starting with $n = 1$. This integer is now known as the **principal quantum number** for the electron.

According to Bohr's model, the radius of the orbit increases as n increases. Thus, the lowest energy state for the electron, the most negative potential energy state for the electron, corresponds to $n = 1$ and the electron being closest to the nucleus. The electron of the hydrogen atom is normally in this lowest energy state and is said to be in its **ground state**. *Energy must be supplied to overcome the attractive force of the proton at the nucleus and move the electron to a higher energy state further from the nucleus.* When the electron is in a higher energy state, the atom is said to be in an **excited state**. Notice, however, from Fig. 7.11 in the text that *the energy difference between successive energy states becomes smaller as n becomes larger.*

Electrons in excited states naturally move back down to lower energy levels and emit energy as they do so. The energy is observed as light that is the source of lines observed in the **emission spectrum** of atoms. When electrons move from states with n greater than 1 to that with $n = 1$ for H atoms, the series of lines is observed in the ultraviolet region and is called the **Lyman series.** When electrons move from states with n greater than 2 to that with $n = 2$ for H atoms, the series of lines is observed in the visible region and is called the **Balmer series**.

In summary, we now realize that the origin of atomic line spectra is the movement of electrons between quantized energy states. The energy of any given line in the spectrum of hydrogen atoms is given by the energy difference between the states, by

$$\begin{aligned} \Delta E &= E_{final} - E_{initial} \\ &= -Rhc(1/n^2_{final} - 1/n^2_{initial}) \qquad (7f, 7.5) \end{aligned}$$

where Rhc is 1312 kJ/mol. *However, Bohr's model is only able to account for the spectra of species having just one electron and the idea of electrons moving about nuclei in paths of fixed radii is no longer considered to be accurate.*

7.4 The Wave Properties of the Electron

The dual wave and particle nature of light was demonstrated by Einstein. *Thus, Louis deBroglie proposed that tiny particles like electrons could also exhibit dual wave and particle properties. In 1925, deBroglie proposed that the wavelength of a free electron of mass m moving with a velocity v could be calculated using*

$$\lambda = \frac{h}{mv} \qquad\qquad (7g, 7.6)$$

Experimental proof of the wave nature of electrons was provided by Davisson and Germer who showed that a beam of electrons was diffracted by the atoms of a thin sheet of metal foil analogous to beams of light waves.

7.5 The Wave Mechanical View of the Atom

Niels Bohr headed a team of physicists who sought to derive a comprehensive theory for the behavior of electrons starting with the particle view of electrons whereas Erwin Schrödinger started with the wave view of electrons. The theory developed by Schrödinger is called **quantum mechanics**, or **wave mechanics**, and is the theory that is used by theoreticians today.

The Uncertainty Principle

Werner Heisenberg concluded that if electrons have wave properties it would be impossible to state both the position and energy of the electron with any degree of certainty. Based on Heisenberg's idea, which we now call the **uncertainty principle**, *Max Born proposed that if we choose to know the energy of an electron in an atom with only a small degree of uncertainty, then we must choose to accept a correspondingly large uncertainty about its position in space about the atom's nucleus.* In practical terms, this means the only thing we can do is calculate the **probability** of finding an electron with a given energy within a given space.

Schrödinger's Model of the Hydrogen Atom and Wave Functions

Schrödinger's wave mechanical approach resulted in a complex mathematical equation that is difficult to solve except in simple cases. The solutions to the equation, called **wave functions**, are chemically important.

Wave functions characterize the electron as a matter wave and are given the Greek letter ψ (psi) as a symbol. The following points concerning wave functions are important:

- The behavior of the electron in an atom is best described as a standing wave, so only certain wave functions are allowed.

- Each allowed wave function ψ corresponds to an allowed energy for the electron. The energy of the electron is therefore quantized. However, quantization enter Schrödinger's theory naturally in contrast to Bohr's theory in which quantization was imposed as a postulate at the start.

- The square of ψ gives the probability of finding the electron within a given region of space. Scientists refer to this as the **electron density** in a given region.

- The matter waves for the allowed energy states are also called **orbitals**.

- To solve Schrödinger's equation for an electron in three-dimensional space, three integer numbers, the quantum numbers n, ℓ and m_ℓ, must be introduced.

Quantum Numbers

Three integer numbers called **quantum numbers** must be introduced to solve Schrödinger's equation for an electron in three-dimensional space. *These numbers have restricted values and are a natural consequence of the theory.*

n, the principal quantum number = 1, 2, 3, ...∞

The principal quantum number n can have any integer value from 1 to infinity. This quantum number is called the principal quantum number because the value of n is the principal factor in determining the energy of the electron. The energy expression obtained for the electron of the hydrogen atom is identical to that obtained by Bohr, namely Expression 7e. Thus, the energy for the electron of the hydrogen atom varies only with the value of n.

The value of n is also a measure of the most probable distance of the electron from the nucleus; the greater the value of n, the more probable that the electron is further from the nucleus. Electrons having the same value of n are said to be in the same **electron shell**.

ℓ, the angular momentum quantum number = 0, 1, 2,...$(n-1)_{max}$

The electrons of a given shell can be grouped into **subshells**, with each subshell characterized by a different value of the quantum number ℓ and by a characteristic shape. *The value of n limits the number of subshells possible for the nth shell because ℓ can be no larger than n-1. Thus, for the nth shell, n different subshells are possible (ℓ = 0 through n-1). Each value of ℓ corresponds to a different subshell and a different orbital type and shape.*

The values of ℓ are usually coded by letters according to the following scheme:

Value of ℓ:	Subshell label:
0	s
1	p
2	d
3	f

The scheme then continues alphabetically, omitting the letter j.

m_ℓ, the magnetic quantum number = 0, ±1, ±2, ±3,....±ℓ_{max}

The number of allowed m_ℓ values equals the number of orbitals found within each subshell. The number of allowed m_ℓ values for each value of ℓ is limited by the fact that m_ℓ can range from +ℓ to -ℓ, with 0 included. Thus, $2\ell+1$ values of m_ℓ are possible for each value of ℓ leading to the following relation:

Type of subshell:	Number of orbitals in the subshell:
s	1
p	3
d	5
f	7

Notice that as ℓ increases by 1, the number of orbitals increases by 2 corresponding to two new values for m_ℓ.

Useful Information from Quantum Numbers

The three quantum numbers n, ℓ and m_ℓ tell us to which shell an electron is assigned (n), to which subshell within the shell (ℓ) and to which orbital within the subshell (m_ℓ), though we can only use m_ℓ values to determine whether two electrons have the same value of m_ℓ and are assigned to the same orbital (because we cannot connect specific values of m_ℓ with specific orbitals). The allowed combinations of quantum numbers for n = 1 through 4 are summarized in Table 7.1 in the text. ✱ Try preparing a "flash card" summarizing the relations between the allowed values of the quantum numbers and a separate "flash card" for each quantum number summarizing the importance of that quantum number.

7.6 The Shapes of Atomic Orbitals

The chemistry of an element is determined by the arrangement of its electrons, especially its outer electrons. Thus, shapes of orbitals are important to understanding the chemistry of the elements.

s Orbitals

The **electron cloud** picture for a 1*s* orbital that is shown in Fig. 7.16a in the text indicates **there is a high probability of finding the 1*s* electron near the nucleus and that the probability of finding the electron decreases rapidly with increasing distance from the nucleus.** This is shown in still another way in Fig. 7.16b in the text where ψ^2 is plotted versus the distance, r, between the nucleus and the electron.

The shape of a 1s orbital is commonly represented as being spherical where the surface of the sphere is taken to be a "boundary surface" that encloses a region in which there is a 90% probability of finding the electron, though it must always be understood that the probability varies considerably within the volume enclosed by the surface of the sphere. Analysis of ψ functions for s orbitals of higher n values results in similar patterns with the size of the s orbital increasing as n increases. However, closer examination also reveals the existence of n-ℓ-1 = n-1 spherical nodes for s orbitals of higher n values.

p Orbitals

There are three p orbitals in each shell having a p subshell, and all have the same basic shape. All p orbitals have one imaginary plane that slices through the nucleus and divides the electron density in half, as shown in Figs. 7.17 and 7.19 in the text. The imaginary plane is called a **nodal plane**, a planar surface on which there is zero possibility of finding the electron. *Thus, unlike s orbitals, there is no probability of finding an electron in a p orbital at the nucleus and plots of ψ^2 necessarily start at zero probability at the nucleus, as shown in Fig. 7.18 in the text.*

The boundary surface diagrams that show an enclosure of 90% probability of finding an electron in a p orbital are described as being dumbbell shaped. There are three p orbitals that lie along the mutually perpendicular x, y and z axes and are labeled as the p_x, p_y and p_z orbitals according to the axis along which they lie. The shapes of the 2p and 3p orbitals are shown in Fig. 7.17 in the text.

d Orbitals

There are five d orbitals in each shell having a d subshell, and following the pattern of ℓ nodal surfaces, each has two nodal surfaces and therefore four regions of electron probability. The d_{xy}, d_{xz} and d_{yz} orbitals lie in the planes formed by the intersections of the x and y, x and z and y and z axes, respectively, and bisect the areas formed by the intersections of these axes, so that the four regions of probability are all 45° off each axis. The four regions of the $d_{x^2-y^2}$ lie along the x and y axes. The final d orbital, the d_{z^2}, has two main regions of electron density along the z axis, but a "donut" of electron density also occurs in the xy-plane. The shapes of the 3d orbitals are shown in Fig. 7.17 in the text.

f Orbitals

The seven f orbitals have three nodal surfaces making them less easily visualized.

7.7   Key Expressions

Wavelength, frequency, speed of electromagnetic radiation expression:

$$\lambda \cdot \nu = c = 2.998 \times 10^8 \text{ m/s}$$

Energy, frequency, wavelength expression:

$$E = h \cdot \nu = \frac{h \cdot c}{\lambda}$$

H atom energy state expression:

$$E_n = \frac{-Rhc}{n^2}$$

H atom energy state difference expression:

$$\Delta E = E_{final} - E_{initial}$$
$$= -Rhc(1/n^2_{final} - 1/n^2_{initial})$$

Wavelength, mass, velocity expression:

$$\lambda = \frac{h}{mv}$$

Quantum number expressions:

n, the principal quantum number = 1, 2, 3, ...∞

ℓ, the angular momentum quantum number = 0, 1, 2,...$(n-1)_{max}$

m_ℓ, the magnetic quantum number = 0, ±1, ±2, ±3,....±ℓ_{max}

Chapter 7

CHAPTER 8

ATOMIC ELECTRON CONFIGURATIONS
and CHEMICAL PERIODICITY

The model of the atom that was developed in Chapter 7 can only be used to describe atoms or ions having a single electron. This chapter extends the model to multielectron species. It also examines factors influencing the sizes of atoms and ions and the ease with which atoms can lose or gain electrons to form ions. The periodic trends in these properties are directly related to the tendency of the elements to form ionic compounds and therefore to the chemistry of the elements.

8.1   Electron Spin

The three quantum numbers n, ℓ and m_ℓ allow us to define an orbital for an electron. However, around 1920, chemists realized that electrons interact with magnetic fields and therefore behave as though they spin about an axis, much like the earth spins or rotates about its axis. It is necessary to understand something about magnetism to be able to understand this, but *we shall see that one more quantum number, the electron spin quantum number, m_s, is required to fully describe electrons*.

Magnetism

In 1600, William Gilbert concluded that the earth acts as a large spherical magnet giving rise to a magnetic field that surrounds the planet. The magnetic field comes out of one end of the earth that is arbitrarily called the "north magnetic pole," N, and loops around toward the "south magnetic pole," S. Furthermore, it can be shown that identical magnetic poles (N-N or S-S) repel each other and opposite magnetic poles (N-S) attract each other. Thus, because the magnetic north pole of a compass points toward the north geographic pole of the earth, we know that the north geographic pole of the earth is actually the magnetic south pole of the earth.

Paramagnetism and Unpaired Electrons

Most substances are repelled by magnets and are said to be **diamagnetic**. However, many metals and other compounds are attracted to a magnetic field and are said to be **paramagnetic.** Still other materials, such as magnetite (Fe_3O_4) and "Alnico" (an Al, Ni and Co alloy) are so strongly attracted to magnets that

---

they are said to be **ferromagnetic**. *It can be shown that these magnetic properties of substances arise from electron spins. In fact, experiments have shown that there are only two spin orientations possible. One orientation is associated with a spin quantum number of m_s = ½ and the other with m_s = -½. Thus, electron spin is quantized.*

Atoms having just one unpaired electron are paramagnetic and exhibit two different interactions with an external magnetic field corresponding to two opposite spins. Thus, when one electron is assigned to an orbital, it can have either spin. However, when two electrons are assigned to the same orbital, as with helium, the magnetic field of the one is canceled by the magnetic field of the other. This can only occur if the electrons possess opposite spins. *Hence, paramagnetism occurs with substances containing atoms or ions having at least one unpaired electron whereas diamagnetism occurs with substances in which all electrons are paired with partners of opposite spin.*

8.2   The Pauli Exclusion Principle

In 1925, Wolfgang Pauli put forth the **Pauli exclusion principle** stating that no two electrons in an atom can have the same set of four quantum numbers (n, ℓ, m_ℓ and m_s). *This means that only two electrons can occupy an orbital and that these electrons must have opposite spins to do so,* because the first three quantum numbers are used to specify an orbital and m_s can only be +½ and -½ corresponding to opposite spins.

This restriction leads to the following subshell capacities:

Type of Subshell:	Electron Capacity:
s	2
p	6
d	10
f	14

The number of orbitals per subshell is thus given by $2\ell+1$, the number of electrons per subshell is given by $2[2\ell+1]$ and the number of electrons per shell is given by $2n^2$, as summarized in Table 8.1 in the text. ✶ Try preparing a "flash card" summarizing the electron capacities of subshells and memorizing it. Also try preparing a second "flash card" summarizing the information of the last sentence and memorizing it.

8.3   Atomic Subshell Energies and Electron Assignments

Electrons can be assigned to shells, subshells and orbitals of multielectron atoms in a manner that makes the total energy of the atom as low as possible. However, we will consider the experimental evidence that exists for the electron configurations of multielectron atoms before we consider the details of making these assignments based on quantum mechanics. We shall see that the quantum mechanical model of the atom gives a reasonably successful account for the experimental observations.

Experimental Evidence for Electron Configurations

In Section 4.3 in the text, you were told that positive ions can be formed by removing electrons from atoms. The energy required to overcome the attractive force of the protons in the nucleus and remove an electron from an atom in the gas phase is called the **ionization energy**. *The second ionization energy of an element is always larger (more endothermic) than the first because the second electron is being removed from a positive ion whereas the first is removed from a neutral atom.*

According to Fig. 8.5 in the text, the ratio of the second ionization energy to the first ionization energy is much higher for Li, Na and K than it is for other elements of the first 36 elements in the periodic table. Furthermore, the ratio is relatively constant for the other elements, though it is slightly higher for B and Al and Ga. *This implies,*

   • the ionization energy for removing the first electron from Li, Na and K is low because these elements begin a new row of the periodic table and the electron that is removed occupies a new shell whereas the ionization energy for removing the second electron is high because the second electron is removed from the $n-1$ inner shell that lies much lower in energy than the outer nth shell,

   • the ratios for the other elements are relatively constant because both electrons are being removed from the same shell, and

   • the ratios for B, Al and Ga are slightly higher than those for the other elements because the second electron for B, Al and Ga must come from a lower energy subshell.

We can now resume our discussion of the currently accepted quantum-mechanical view of atoms and see whether these experimentally-based inferences agree with the results of quantum mechanics.

Order of Subshell Energies and Assignments

Quantum theory tells us the energy of the electron of the hydrogen atom only depends on the value of n (7e,7.4). However, the ionization energy plot of Fig. 8.5 in the text and the experimentally determined order of subshell filling shown in Fig. 8.6 in the text both indicate that *the subshell energies of multielectron atoms depend on both n and ℓ. Thus, unlike the case with H atoms, the subshells of any given shell of multielectron atoms have different energies.* For example, for H atoms $3s = 3p = 3d$ in energy, but for atoms of other elements $3s < 3p < 3d$ is observed.

The subshell energy order of Fig. 8.6 in the text and the experimentally determined electron configurations of the elements lead to *two general guidelines that can be used to predict the electron configurations of the elements. They are:*

- Assign electrons to subshells in order of increasing "$n + \ell$" value, as shown by the diagonals in Fig. 8.7 in the text.

- Assign electrons to the subshell of lower n when two subshells have the same "$n + \ell$" value, as shown along any given diagonal in Fig. 8.7 in the text.

Notice that this involves:

- Assigning electrons to subshells having the same ℓ value in an order of increasing n value. For example, when $\ell = 1$, the subshell order of filling is $2p$, $3p$, $4p$, etc. *When ℓ is constant, subshells fill in an order of increasing n.*

- Assigning electrons to the subshells of any given n value shell in an order of increasing ℓ value. For example, when $n = 4$, the subshell order of filling is $4s$, $4p$, $4d$ and $4f$. *When n is constant, subshells fill in an order of increasing ℓ.*

* Try preparing a "flash card" showing the subshell order of filling that is given in Fig. 8.7 in the text and memorizing it.

Effective Nuclear Charge, Z*

The order in which electrons are assigned to subshells of atoms and many atomic properties can be rationalized by the concept of *effective nuclear charge. The regions occupied by electrons in the various subshells of an atom are not mutually exclusive,* as shown in Figs. 7.18 and 8.8 in the text. *When the probability pattern of an outer subshell overlaps that of an inner subshell, we say that the outer subshell penetrates the region occupied by the inner subshell. In general, the order of penetration is s > p > d > f and this is why the order of subshell filling is s, p, d and then f in any shell having these subshells available.* The charge felt by any electron at any given distance from the nucleus is called the **shielded nuclear charge**, and the average shielded charge felt by the electron based on its total probability pattern is called the **effective nuclear charge** (Z^*). *Penetration increases the effective nuclear charge experienced by electrons and lowers their potential energy. This is why the observed order of subshell filling is s, p, d, f, etc. within each shell.*

8.4   Atomic Electron Configurations

The electron configurations of the elements through element 109 are given in Table 8.2 in the text. These are the configurations which give the lowest potential energies and are called the **ground state electron configurations**.

Electron Configurations of the Main Group Elements

The electron configuration of an element can be written using **orbital box diagrams** or **spectroscopic notation,** as shown for the first 10 elements in Table 8.3 in the text. *With orbital box diagrams, one box is used for each orbital and an appropriate number of boxes is used for each subshell. Upward and downward arrows are used to represent the electrons, with the custom of the text being to first place one upward arrow in each box within a given subshell before placing a second and downward arrow in any of the boxes for that subshell.* See the orbital box diagrams given for C through Ne in Table 8.3 in the text, and note that this procedure is in accord with **Hund's rule**, which states that the most stable arrangement of electrons is that with the maximum number of unpaired electrons, all with the same spin direction.

With the spectroscopic notation, the symbolism

$$\text{(number)(letter)}^{\text{superscript number}}$$

is used where the number represents the n value of the subshell being filled, the letter represents the ℓ value of the subshell *being filled and the superscript number gives the number of electrons in that subshell.* Thus, $2p^5$ represents 5 electrons in the p subshell of the second ($n = 2$) shell.

In writing electron configurations for elements, electrons are assigned to subshells following the subshell filling order of Fig. 8.7 of the text, the guideline of Hund's rule, the electron capacities of subshells that were developed using the allowed combinations of quantum numbers and the Pauli exclusion principle (8.2). The total number of electrons to be used for neutral atoms is given by the atomic number of the element, because neutral atoms necessarily have equal numbers of protons and electrons. The resulting electron configurations are often written in an abbreviated form which makes use of the preceding **noble gas configuration**, as with $[\text{Ne}]3s^1$ to represent the $1s^2 2s^2 2p^6 3s^1$ electron configuration of Na.

The electrons that are included in the noble gas configuration are often referred to as the **core electrons** of the atoms, and those beyond the core are referred to as the **valence electrons** of the atom, the electrons that determine the chemical properties of the atom (3.3). *Using either means of representation, we notice that all the elements of a given group of the main group elements have the same number of valence electrons and thus the same general valence electron configuration. The number of valence electrons for the atoms of any given main group element, any A group element, is always equal to its group number.* Thus, atoms of oxygen and sulfur from Group VIA both have six valence electrons, and their respective electron configurations are written as $[\text{He}]2s^2 2p^4$ and $[\text{Ne}]3s^2 3p^4$.

Electron Configurations for the Transition Elements

Elements whose atoms are filling d subshells are often referred to as the **transition elements**. Those for which f subshells are filling are sometimes called the *inner transition elements*, or more commonly, the **lanthanides** (filling $4f$ orbitals) and **actinides** (filling $5f$ orbitals). *Exceptions to the general filling order of Fig. 8.7 of the text occur in these regions of the periodic table because the subshells are of similar energies.* For example, the expected electron configurations for Cr and Cu of the *first transition series* are $[\text{Ar}]4s^2 3d^4$ and $[\text{Ar}]4s^2 3d^9$, respectively, and the observed electron configurations are $[\text{Ar}]4s^1 3d^5$ and $[\text{Ar}]4s^1 3d^{10}$. However, chemists are usually concerned with the chemistry of the elements and thus the positive ions that are formed by the metals. *The differences between the predicted and observed electron configurations for neutral atoms usually have little or no effect on the chemistry of the elements, because the electron configurations of ions of these elements show no such "anomalies."*

8.5   Electron Configurations of Ions

To form a positive ion, a cation, of an element, one or more electrons are removed from the electron shell of highest n. If there is a choice of subshells within the highest n shell, the electrons of maximum ℓ are removed first. This leads to the following changes in electron configurations:

$$Mg[1s^2 2s^2 2p^6 3s^2] \rightarrow Mg^{2+}[1s^2 2s^2 2p^6] + 2e^-$$

$$Fe[Ar]3d^6 4s^2 \quad\;\; \rightarrow Fe^{2+}[Ar]3d^6 + 2e^-$$

$$Fe[Ar]3d^6 4s^2 \quad\;\; \rightarrow Fe^{3+}[Ar]3d^5 + 3e^-$$

$$Sn[Kr]4d^{10} 5s^2 5p^2 \rightarrow Sn^{2+}[Kr]4d^{10} 5s^2 + 2e^-$$

$$Sn[Kr]4d^{10} 5s^2 5p^2 \rightarrow Sn^{4+}[Kr]4d^{10} + 4e^-$$

The magnetic properties of the ions formed can be used to confirm these changes, because the extent of paramagnetism is related to the number of unpaired electrons. For example, it can be shown that Fe^{3+} is paramagnetic to the extent of having 5 unpaired electrons, whereas removing electrons from the $3d$ before the $4s$ would have yielded just 3 unpaired electrons as $[Ar]3d^3 4s^2$. ✶ Try preparing a "flash card" summarizing the order of removal of electrons to form cations and memorizing it.

8.6   Atomic Properties and Periodic Trends

The objective of this section is to show how some physical properties of the elements are related to their electron configurations and therefore vary in systematic patterns in the periodic table. The properties of atomic size, ionization energy, electron affinity and ion sizes will be considered.

Atomic Size

Atoms do not have finite, definitive boundaries where the probabilities of finding electrons go to zero. However, internuclear distances can be determined experimentally, and estimates of atomic radii can be obtained by taking one-half the internuclear distance between like atoms in diatomic molecules, such as H_2 and Cl_2. The radii obtained for H and Cl can then be added together and compared to the internuclear distance in HCl, and it can be shown that there is reasonable agreement between the calculated sum and the experimental value. The radius of H can then be used with the experimental value for the H-O internuclear distance in H_2O to obtain a value for the radius of O. The radius of Cl can likewise be used with the experimental value for the C-Cl internuclear distance in CCl_4 to obtain a value for the radius of C. The radius of C can then be used with the experimental value for the C-Br internuclear distance in CBr_4 to obtain a value for the radius of Br. Using this method and other methods has led to the set of atomic radii for atoms of main group elements that is given in Fig. 8.10 in the text.

For the main group, s and p block, elements, atomic radii decrease going across a period and increase going down a group in the periodic table.

- Radii decrease going across a period, because electrons are added to valence orbitals having the same n value and the increase in the number of protons at the nucleus causes an increase in effective nuclear charge, $Z*$.

- Radii increase from top to bottom within a main group, because electrons are assigned to orbitals in shells that are further from the nucleus.

For the transition elements, the trend is less systematic.

- Radii initially decrease slightly proceeding across a given transition series and then begin to increase as pairing within the d subshell occurs, because there is repulsion between the paired d electrons themselves and the repulsion between the increasing d electrons and the outer s electrons also increases.

- Radii of transition elements increase on moving from elements in the 4th period to corresponding elements in the 5th period, but corresponding elements in the 5th and 6th periods have nearly identical radii. The latter is due to the fact that the poor shielding $4f$ subshell fills between the 5th and 6th period transition elements leading to substantial increases in effective nuclear charge, $Z*$.

Ionization Energy

In 8.3, **ionization energy** was defined as the energy that is required to remove an electron in the gas phase.

$$A(g) \rightarrow A^+(g) + e^- \quad \Delta E \equiv \text{ionization energy, } IE$$

Energy is always required to overcome the attraction of the

effective positive charge of the nucleus, Z*, to remove an
electron from an atom. Thus, the sign of the ionization energy
is always positive in accord with thermodynamic convention.
Successive ionizations become increasingly more positive because
electrons are being removed from increasingly more positive
ions. The increase becomes especially great when removing an
electron from a shell of lower n.

For the main group elements, first ionization energies generally
increase going across a period and decrease going down a group
in the periodic table.

- Ionization energies generally increase going across a
  period, because Z^* increases and size decreases. There are,
  however, two exceptions to this general trend of increasing
  first ionization energies:

  - There is a decrease from Group IIA to Group IIIA,
    because the electron that is removed from the Group
    IIA element comes from an ns subshell whereas the
    electron that is removed from the Group IIIA element
    comes from the less penetrating and inherently higher
    energy np subshell.

  - There is a decrease from Group VA to Group VIA,
    because the repulsion caused by pairing in the np^4
    configuration of the Group VIA element overcomes the
    increase in Z^* and decrease in size from Group VA to
    Group VIA. (However, the increase in Z^* and decrease
    in size from Group VIA to Group VIII overcomes the
    effects of pairing in elements in these Groups, and
    the normal trend of increasing *IE* resumes.)

- Ionization energies decrease going from top to bottom
  within a group, because the increase in size associated
  with filling new shells overcomes the increase in Z^*.

Electron Affinity

The **electron affinity, *EA,*** of an atom is the energy change that
occurs when an electron is added to an atom in the gas phase.

$$A(g) + e^- \rightarrow A^-(g) \qquad \Delta E \equiv \text{electron affinity, } EA$$

Energy is usually released when the first electron is added to
an incomplete subshell of an atom of a main group element.
Periodic trends in electron affinity are closely related to
those for ionization energies, because the same factors, namely
Z* and size, govern the attractions for electrons and therefore

---

affect both ionization energies and electron affinities. An element with a high endothermic (positive) ionization energy generally has a high exothermic (negative) electron affinity.

For the main group elements, first electron affinities generally increase going across a period and decrease going down a group in the periodic table.

• Electron affinities generally increase going across a period, because Z^* increases and size decreases. There are, however, three exceptions to this general trend of increasing first electron affinities:

   • There is a decrease from Group IA to IIA, because the elements of Group IIA already have a completed s subshell, and the additional electron must be assigned to the higher energy p subshell. Indeed, the electron affinities shown for Be and Mg in Fig. 8.14 in the text are zero, indicating the members of Group IIA have no affinity for an additional electron.

   • There is a decrease from Group IVA to Group VA, because the electron that is added to the p^3 configuration of the Group VA element must pair with one of these electrons. The increase in Z^* and decrease in size from Group IVA to Group VA does not overcome the repulsion associated with the p^4 configuration, and members of Group VA have no affinity for an additional electron.

   • The members of Group VIII have completed s and p subshells and therefore have no affinity for an additional electron, because the added electron would be assigned to the s subshell of the next shell.

• Electron affinities generally decrease going from top to bottom within a group, because the increase in size associated with filling new shells overcomes the increase in Z^*. There is one exception to this general trend.

   • The electron affinities of members of the second period p block elements (B, C, N, O and F) are lower than those for the corresponding third period elements (Al, Si, P, S and Cl), because adding an electron to the very small atoms of the second period elements causes considerable repulsion and therefore offsets the greater attraction that is associated with placing the electron closer to the nucleus.

Ion Sizes

• The periodic trend for sizes of ions parallels the trend for the sizes of neutral atoms within groups: the sizes of the positive or negative ions increases from top to bottom within a group, as shown in Fig. 8.15 in the text.

• The sizes of cations are always smaller than the neutral atoms from which they are derived, because the attractive force of the nuclear charge acts more effectively on fewer electrons.

• The sizes of anions are always larger than the neutral atoms from which they are derived, because the attractive force of the nuclear charge acts less effectively on more electrons and there is also an increase in electron-electron repulsions.

• The sizes of ions having the same number of electrons, of ions which are **isoelectronic**, decreases with increasing nuclear charge, because the increasing nuclear charge acts more effectively on the constant number of electrons.

8.7 Chemical Reactions and Periodic Properties

This section discusses the reactions of metals with the halogens (Group VIIA) in terms of the atomic properties and periodic trends that have been discussed in this chapter.

Energy of Ion Pair Formation

Ionic compounds usually contain metal cations and nonmetal anions; the only common exception involves ionic compounds containing NH_4^+ cations. One measure of the strength of attraction between two oppositely charged ions is the enthalpy of formation of a cation-anion pair, with everything in the gas phase. The greater the attraction between the ions, the more negative the value of $\Delta H_{formation}$.

The value of $\Delta H_{formation}$ can be estimated by using an equation that is derived from Coulomb's law, which is the law that describes the energy of interaction between charged particles (page 119 in the text).

$$\Delta H_{formation} \propto -N(n_+ n_- / d) \qquad \text{(8a, 8.1)}$$

where n_+ is the number of positive charges on the cation, n_- is *the number of negative charges on the anion, d is the distance between the ion centers and N is Avogadro's number, so that the energy is estimated for the formation of one mole of ion pairs. The greater the charges on the ions and the smaller the distance between the centers of the ions, the greater the attraction and the more negative (more exothermic) the $\Delta H_{formation}$.*

The enthalpies of formation for the gaseous alkali metal halide ion pairs are shown in Fig. 8.17 in the text. *Notice that*

- the negative $\Delta H_{formation}$ values for the formation of the alkali metal halides of any given halogen decrease in the order LiX > NaX > KX, because the increasing cation size from Li^+ to K^+ causes d to increase and $\Delta H_{formation}$ to decrease in accord with Expression 8a.

- the negative $\Delta H_{formation}$ values for the formation of the halides of any given alkali metal decrease in the order MF > MCl > MBr > MI because the increasing anion size from F^- to I^- causes d to increase and $\Delta H_{formation}$ to decrease in accord with Expression 8a.

Lattice Energy

Although the formation of ion pairs is useful for illustrating periodic trends, *it must be remembered that ionic compounds are composed of multitudes of cations and anions arranged into a crystalline lattice*. The enthalpy change for bringing gaseous cations and anions together to form one mole of an ionic solid is called the **lattice energy**, or **lattice enthalpy.** Though many cation-anion attractions and many cation-cation and anion-anion repulsions contribute to the lattice energy of the alkali metal halides, the trends are still those observed with the formation of ion pairs. *The greater the charges on the ions and the smaller the distance between the centers of the ions, the greater the attraction and the more negative the lattice energy, the more exothermic the lattice formation*. Lattice energies can be calculated using an equation derived from Expression 8a.

The enthalpy of formation of an ionic compound can be broken down into a series of steps that includes the lattice formation and analyzed in terms of these steps. The approach that is used is called a **Born-Haber cycle**, named for Max Born and Fritz Haber who were prominent German scientists, and is based on Hess's Law that was first introduced in Section 6.6 in the text. *The total enthalpy change for the formation of the gaseous cations and anions is always positive in these cycles. It is the negative enthalpy change for lattice formation which pays the cost of ion formation and leads to the exothermic ΔH_f^o values for ionic compounds, according to this model.*

Why Do Binary Ionic Compounds of Main Group Elements Usually Contain Ions With Noble Gas Configurations?

Binary ionic compounds of main group elements almost always contain ions with noble gas configurations; the only common exceptions are the peroxides and superoxides of elements of Groups IA and IIA elements. Born-Haber cycles can be used to rationalize this observation, though we must keep in mind that there is no experimental evidence that suggests atoms actually go through these steps when forming ionic compounds. The Born-Haber cycle merely represents an energetically equivalent alternative to direct reaction, based on the state function nature of enthalpy.

There are three important points to consider when rationalizing the charges on ions in binary ionic compounds and the corresponding electron configurations of these ions:

• The removal of successive electrons from an atom of a metal becomes increasingly more endothermic. This energy is repaid by the increase in lattice energy that is gained by bringing cations of higher charge and smaller size together with any given anion, provided the removal of electrons does not cause the cation to go beyond a noble gas configuration. This restriction is due to the fact that there are large increases in ionization energies whenever metals go beyond noble gas configurations, because the energy that is required to remove an electron from a shell of lower n value is great. This energy is not repaid by the increase in lattice energy that would be gained by bringing cations of still higher charge and smaller size together with any given anion.

• The addition of two or more electrons to an atom is endothermic, because second and successive electrons are repelled by the negative charge of the anion formed by adding the first electron. This energy is repaid by the increase in lattice energy that is gained by bringing anions of higher charge together with any given cation, provided the number of electrons added does not cause the anion to go beyond a noble gas configuration. This restriction is due to the fact that the noble gas configuration has completely filled subshells, and that the added electron would need to enter a shell of higher n value. Such an addition is likely to require energy rather than release energy and thereby offset any energy advantage to be gained by bringing together ions of higher charge.

The noble gases have the highest ionization energies and lowest electron affinities within their periods, and thus do not form

ions. The tendency of atoms to achieve **noble gas configurations** is often referred to as tendency to achieve an **electron octet**, because the noble gases below helium all have ns^2np^6 valence electron configurations.

8.8   Key Expressions

Gas Phase Ion Pair Enthalpy of Formation Expression:

$$\Delta H_{formation} \propto -N(n_+n_-/d)$$

8.9   Hints for Answering Questions and Solving Problems

Writing Electron Configurations of Atoms:

The arrangement of the elements in the periodic table follows the subshell order of filling given in Fig. 8.7 in the text. Thus, Fig. 8.9 in the text can be used with a traditional representation of the periodic table to predict the electron configurations of atoms. Se, for example, is in the fourth period, and the preceding noble gas is Ar. Thus, the electron configuration of Se is $[Ar]4s^23d^{10}4p^4$, because it is the 4th element in the 4p region. * Try preparing a "flash card" having the form of Fig. 8.9 and memorizing it. Also try using this method to predict the electron configurations of the main group and first transition series elements and comparing your answers to the configurations given in Table 8.2 of the text.

Predicting Periodic Properties of Atoms:

• Atomic radii generally decrease going across a period and increase going down a group.

• Ionization energies generally increase going across a period and decrease going down a group.

• Electron affinities generally increase going across a period and decrease going down a group, though the trend going across does not include the members of Group VIII.

• Ion sizes generally increase going down a group.

- Ion sizes of isoelectronic cations decrease going across a period whereas ion sizes of isoelectronic anions decrease going across a period.

* Try preparing a "flash card" summarizing the general periodic trends for atomic radii, ionization energies and electron affinities and memorizing it. Try preparing a separate "flash card" summarizing the general trends in ion sizes and memorizing it.

Chapter 8

CHAPTER 9

BONDING and MOLECULAR STRUCTURE: FUNDAMENTAL CONCEPTS

This chapter describes the use of outer electrons, called valence electrons, in the formation of covalent bonds involving shared pairs of electrons. The shared pairs of electrons are shown in Lewis structures, which also show the presence of any nonbonding or lone pairs of electrons in the valence shells of the bonded atoms. Lewis structures can be used to rationalize bond orders, lengths and strengths and molecular shapes.

Molecular shapes are very important, because the properties of covalent molecules are related to their molecular shapes. These properties determine, for example, whether the covalent substances are gases, liquids or solids at room temperature and pressure and whether the covalent substances are soluble in water or insoluble in water. Hence, the material in this chapter should be mastered to be used in the study of later chapters.

9.1 Valence Electrons

The electrons of an atom can be divided into two groups, the **valence electrons**, which participate in chemical bonding with other atoms, and the **core electrons**, which do not.

> • *The valence electrons for the main group elements are those in the outermost s and p subshells.* The core electrons are those in the inner shells. For Br having the electron configuration $[Ar]4s^23d^{10}4p^5$, the valence electrons are the $4s^24p^5$ electrons, and the core electrons are those represented by $[Ar]3d^{10}$.

> • *The valence electrons for the transition elements are those in the outermost ns and (n - 1)d subshells.* The core electrons are those in the inner subshells. For Co having the electron configuration $[Ar]4s^23d^7$, the valence electrons are the $4s^23d^7$ electrons, and the core electrons are those represented by [Ar].

The number of valence electrons for atoms of main group elements is always equal to the group number (8.7). *The fact that each element in a main group has the same number of valence electrons accounts for the similar chemical properties within groups.*

Early in this century G.N. Lewis assumed each noble gas had a completely filled outermost shell, which he regarded as a stable configuration, because of the lack of chemical reactivity of these elements. Since each noble gas below helium has a ns^2np^6 valence configuration, this observation became known as the **octet rule**. *Lewis furthermore placed dots around the symbol of an element to represent the valence electrons of an atom, placing one electron around each of the four sides before placing two around any given side*. The valence electrons of atoms of the noble gases below helium were therefore assigned as an octet of four pairs of electrons, and *atoms of other main group elements were similarly assigned a number of valence electrons that was equal to their group number*.

Lewis believed atoms acquired noble gas configurations through chemical reactions. We now know there are numerous exceptions to the so-called octet rule and should only consider it to be a guideline. Nevertheless, it remains useful for predicting the results of most common reactions and will be used to do so.

9.2   Chemical Bond Formation

Chemical bond formation can be described in terms of the attraction between opposite electrical charges and the repulsion between like electrical charges (3.4). For example, when two hydrogen atoms approach each other, the nuclear charge of the one atom attracts the electron of the other atom and vice versa causing the electron clouds of the two atoms to overlap one another. The highest region of electron probability comes in the overlap region between the nuclei, and the attraction between the protons and electrons causes a decrease in potential energy, as shown in Figs. 9.1 and 9.2 in the text. However, at short distances of separation, appreciable proton-proton and electron-electron repulsion occurs causing an increase in potential energy and an overall potential energy minimum. The internuclear distance of separation at the potential energy minimum for H_2 occurs at 74 pm, the equilibrium bond distance for H_2.

The bond between the H atoms in H_2 involves the sharing of electrons and is called a **covalent bond**. *Covalent bonding generally occurs between nonmetals, between elements from the far right hand side of the periodic table.* On the other hand, *ionic bonding involves the transfer of electrons from a metal having a low ionization energy and low electron affinity to a nonmetal having a high ionization energy and high electron affinity*, as in the formation of NaCl (8.7). *However, covalent bonding involving equal sharing of electrons and ionic bonding involving complete transfer of electrons represent the extreme cases of bonding between atoms. The bonding in most compounds is* *actually intermediate between these two limiting cases.*

9.3   Covalent Bonding

The emphasis in this chapter is on bonding involving the sharing of electrons, covalent bonding. In this Section, we will consider the use of Lewis structures to illustrate the sharing of electrons.

Lewis Electron Dot Structures

Lewis electron dot structures can be used to show the distribution of the valence electrons of atoms in molecules. *Lewis structures can be drawn by starting with the Lewis dot symbols for the individual atoms and arranging the valence electrons until each atom has a noble gas configuration*. Examples of Lewis structures by periodic groups are given in Table 9.2 in the text. Notice that *shared pairs of electrons are considered to be part of the valence shells of both atoms*. Also notice that *shared pairs of electrons can be represented by a pair of dots or by a line between the symbols of the two elements*. The shared pairs of electrons are called **bonding** or **bond pairs**, and the other pairs of electrons shown separately in the valence shells of the atoms are called **nonbonding** or **lone pairs**.

Atoms of C, N and O from the second period and atoms of S from the third period commonly form double and triple bonds involving the sharing of two and three pairs of electrons, respectively, to achieve an octet. However, S only forms multiple bonds with the smaller atoms of C, N and O from the second period and does not form stable multiple bonds with itself or other atoms from the third period and below to achieve an octet. *One line is used to represent each shared pair of a multiple bond in a Lewis structure.*

Drawing Lewis Structures

The following guidelines can be used to draw acceptable Lewis structures:

   • Arrange the atoms of the molecule or ion in an acceptable manner. When doing so, realize that the:

      • H atoms can only bond to one other atom and are therefore never central atoms; H atoms are always terminal atoms.

• atom other than H having the lowest electron affinity is usually the central atom.

• H atoms of oxoacids are usually attached to O atoms rather than the central atom. Thus, the formula of sulfuric acid, H_2SO_4, should be rewritten as $SO_2(OH)_2$ when drawing its Lewis structure.

• most symmetrical structures are usually favored for molecules and ions having more than one central atom. Thus, the formulas N_2H_4, C_2H_6 and $P_2O_7^{4-}$ should be rewritten as H_2NNH_2, H_3CCH_3 and $O_3POPO_3^{4-}$, respectively, when drawing their Lewis structures.

• Calculate the total number of valence electrons in the molecule or ion by adding the valence electrons of each atom and adjusting for the net charge, if any.

  • For positive ions, subtract one electron for each unit of positive charge.

  • For negative ions, add one electron for each unit of negative charge.

• Place one pair of electrons between each set of bonded atoms to make a **single bond**.

• Place lone pairs around each of the attached atoms, except H, to satisfy the octet rule.

• If there are electrons remaining, place them around the central atom.

  • Notice that if the central atom is from the third or higher period, it can accommodate more than four pairs of electrons.

• If the central atom is still not surrounded by an octet of electrons, move a lone pair from a terminal atom to form another bond between the central atom and that particular terminal atom. Use the guideline that says that C, N, O and S can form multiple bonds to achieve an octet.

The application of these guidelines is illustrated in Ex. 9.1 in the text and the consideration of NO_2^+ which precedes it. Also notice from the discussion of the text how ***molecules and ions having the same number of atoms and same number of valence*** *electrons have the same Lewis structures*. Species having the same number of electrons are isoelectronic (8.6). ✳ Try practicing following the guidelines for writing Lewis structures until you are so familiar with them that you do not have to take too long to think about them.

Resonance Structures

In some cases it is impossible to draw a Lewis structure that is compatible with the observed properties of the molecule or ion. For example, the bond lengths in ozone, O_3, are intermediate between the bond lengths expected for single and double bonds, and there is no way to draw a Lewis structure to convey this. Hence, Linus Pauling proposed the theory of **resonance** to account for this. Pauling suggested these species are actually composites of contributing structures called **resonance structures.** *Resonance structures differ in the number of bond pairs shown between given pairs of atoms and often occur when there are equivalent ways of writing double bonds*, as shown with O_3, CO_3^{2-} and $CH_3CO_2^-$ in the text.

It is conventional to connect contributing resonance structures with a double-headed arrow, ↔, to indicate the actual structure is a composite or **resonance hybrid** of the contributing structures. However, this practice can be mistakenly interpreted to mean that electrons move back and forth between atoms to generate the contributing structures. They do not, and perhaps the following analogy will help you better understand the concept of resonance. The analogy is often made that the resonance hybrid is like a mule, which is a genetic hybrid between a horse and a donkey. The mule is a mule and does not "resonate" back and forth between being a horse and a donkey. Likewise, *molecules do not "resonate" back and forth between contributing resonance forms. Resonance forms do not adequately represent the actual molecule which has its own unique structure. Resonance is merely a human invention that is intended to address the limitations of drawing Lewis structures.*

Exceptions to the Octet Rule

Compounds in Which an Atom Has Fewer Than Eight Valence Electrons:

Hydrogen can only accommodate two electrons in its valence shell, and *beryllium and boron commonly form compounds in which they are surrounded by just four and six electrons, respectively. Thus, you should generally avoid forming multiple bonds to achieve octets for Be and B as central atoms when drawing Lewis structures.*

Compounds containing atoms other than H with fewer than eight valence electrons tend to react with molecules having lone pairs of electrons to form coordinate covalent bonds. A **coordinate covalent bond** occurs whenever the bonding pair originates with just one of the bonded atoms. For example, BF_3 readily reacts with $:NH_3$ to form $F_3B:NH_3$ in which N furnishes both electrons for the boron to nitrogen covalent bond.

Compounds in Which an Atom Has More Than Eight Valence Electrons:

If the central atom of a molecule is from the third or higher period, it has access to d orbitals and can be surrounded by more than four valence pairs of electrons. Thus, members of these periods can form compounds that cannot be formed by the top members of their respective groups. For example, PCl_5 and SF_6 exist whereas NCl_5 and OF_6 do not.

Compounds with an Odd Number of Electrons:

A few stable molecules contain an odd number of valence electrons and are called **free radicals** because of the presence of unpaired electrons. Examples include NO (11 valence electrons), NO_2 (17 valence electrons and ClO_2 (19 valence electrons). *Free radicals can combine with themselves to form dimers*, such as NO with NO to form N_2O_2 and NO_2 with NO_2 to form N_2O_4, *or react with other molecules to produce new free radicals.* The latter type of reactions play an important role in the chemistry of air pollution.

Oxygen:

Oxygen has an even number of valence electrons (12) but experimental evidence indicates it has both a double bond and two unpaired electrons. *It is not possible to draw a Lewis structure for O_2 which agrees with these experimental results, but chemists can account for these properties of O_2 using the molecular orbital theory of covalent bonding* (10.2).

9.4 Bond Properties

Bond Order

The number of bonding electron pairs shared by two atoms is called the **bond order** of the bond. The distance between bonded atoms (the bond length) and the energy that is required to separate bonded atoms (the bond strength) is related to the bond order. *The higher the bond order between any two bonded atoms, the shorter and stronger the bond, because the higher the* *electron density between the bonded atoms the closer the nuclei can come together and the stronger the attraction between the nuclear charge and the bonding electrons.*

Fractional bond orders occur with molecules and ions having resonance structures. For example, there are two equivalent resonance structures for O_3, and the average bond order is 1.50. Similarly there are three equivalent resonance structures for CO_3^{2-}, and the average bond order is 1.33.

Bond Length

The distance between the nuclei of bonded atoms is called the **bond length**. Average bond lengths between various atoms are given in Table 9.4 in the text. *Notice that:*

- when bonds of the same order are compared, the larger the atoms involved, the longer the bond length. Thus, relative bond lengths can be predicted on the basis of the periodic positions of atoms (Recall Fig. 8.10 in the text). For example, bond length decreases in the order C-C, C-N and C-O, because size decreases from C to O. Similarly, bond length increases in the order Si-F, Si-Cl, Si-Br and Si-I, because size increases from F to I.

- when bonds between like atoms are compared, the higher the bond order, the shorter the bond.

Bond Energy

The enthalpy change for breaking a bond in a molecule with the reactants and products both in the gas phase is called the **bond dissociation energy, D.** *The process of breaking a bond involves overcoming the attractive forces between the nuclei of the bonded atoms and the bonding electrons and is therefore always an endothermic process. The magnitude of energy that is released when forming the same bond in the gas phase is the same, but the process is exothermic, and the sign of the enthalpy change is negative.* Average bond energies between various atoms are given in Table 9.5 in the text. *Notice that*:

- when bonds between like atoms are compared, the higher the bond order, the higher the bond energy.

In gas phase reactions between molecules, original bonds are broken in the reactants and new bonds are formed in the products. If the total energy released when new bonds are formed exceeds the energy that is required to break the original bonds, the overall reaction is exothermic. If the opposite is true, the overall reaction is endothermic.

When using bond energies to calculate the enthalpy change for a gas phase reaction, it is necessary to sum the energies of all the bonds broken and subtract the sum of the energies of all the bonds formed. Thus, we obtain

$$\Delta H^{\circ}_{rxn} = \Sigma D(\text{bonds broken}) - \Sigma D(\text{bonds formed}) \quad (9a)$$

and for the reaction, $CH_4(g) + 2Cl_2(g) \rightarrow CH_2Cl_2(g) + 2HCl(g)$, this means

$\Sigma D(\text{bonds broken}) = 2\text{C-H} + 2\text{Cl-Cl}$

and

$\Sigma D(\text{bonds formed}) = 2\text{C-Cl} + 2\text{H-Cl}$

Perhaps you noticed that the form of Expression 9a differs from the form of Expression involving enthalpies of formation. This is due to the fact that bond energies are always tabulated as endothermic quantities associated with bond breaking.

Bond Polarity and Electronegativity

In Section 8.6 we learned that atoms differ in their abilities to hold onto their valence electrons and to take on an additional electron. This situation carries over into bonds in molecules where unlike atoms differ in their ability to attract shared pairs of electrons. The bonded atom that exerts the stronger attraction for the shared pairs of electrons becomes partially negative, and the other becomes partially positive creating a **polar covalent bond.** *This polarity is indicated by placing a δ_+ symbol by the partially positive atom and a δ_- symbol by the partially negative atom, where δ is the Greek letter "delta."* When shared pairs of electrons are shared equally, as with like atoms in diatomic and other molecules, the bond is **nonpolar covalent.**

In the 1930's, Linus Pauling put forth a scale of electronegativity values to account for the differing tendencies of atoms to attract shared pairs of electrons. The **electronegativity (χ)** value of an element is a measure of the ability of an atom to attract shared electrons to itself. The electronegativity values given in Fig. 9.7 in the text can be used to a) decide if a bond is polar, b) determine which end of a polar bond is positive and which is negative and c) compare the polarity of a bond to that of other bonds. Notice that the *atoms in the lower left side of the periodic table have the lowest electronegativity values whereas those in the upper right have the highest electronegativity values, with fluorine having the highest electronegativity value of any element.* Also notice that *electronegativity values tend to increase in all directions towards fluorine.*

The vastly differing ability of atoms of metals and nonmetals to attract shared electrons is one reason why metallic elements from the left side of the periodic table react with nonmetals from the right side of the periodic table to form ionic compounds. On the other hand, the similarities in electronegativity values of nonmetals is one reason why nonmetals tend to react with each other to form molecular compounds with polar covalent bonds.

Oxidation Numbers and Atom Formal Charges

Oxidation Numbers:

The **oxidation number** of an atom is the charge the atom would have if all of its bonding pairs of electrons were completely transferred to the most electronegative partners in its bonds; that is, if all of its bonds were ionic. Atoms can have relatively high positive oxidation numbers. However, since electrical charges greater than ±3 are highly unlikely, this tells us that oxidation numbers usually do not represent the real charges on atoms. In fact, they only represent real charges for monatomic ions. Nevertheless, oxidation numbers indicate the positively and negatively charged atoms in molecules and thus help us to understand how atoms in molecules react to form new bonds.

Atom Formal Charges:

Oxidation numbers are assigned by assigning shared electrons to the more electronegative atoms of bond pairs and do not give realistic estimates of charges of atoms in molecules. More reasonable estimates of atom charges can be made by assuming that each bond pair is shared equally by bonded atoms, and these charges are called **formal charges**. *The formula that is used to calculate formal charges is:*

atom formal   = group number - number of lone pair electrons
   charge   - ½(number of bonding electrons)   (9b)

The application of this equation is illustrated on pages 427 and 428 and in Ex. 9.6 in the text. Notice that **equivalent formal charges are obtained with equivalent resonance structures but not with nonequivalent resonance forms. Thus, atom formal charges can be used with several guidelines to determine which of two or more nonequivalent resonance structures is more likely. These guidelines are:**

   • Atoms in molecules or ions should have formal charges as small as possible. (This is the so-called *principle of electroneutrality*.)

• Negative formal charges should reside on the most electronegative atoms.

• Like formal charges on adjacent atoms should be avoided, because structures having like formal charges on adjacent atoms are not likely to be stable.

In using the guideline to minimize formal charges, it is common to draw Lewis structures with multiple bonds that place more than four pairs of electrons about atoms from the third period or below, atoms such as S and P. This is acceptable, because these atoms have access to vacant d orbitals having the same principal quantum number as their s and p valence orbitals and are known to form molecules and ions in which they have more than an octet in their valence shell.

Atom Partial Charges

Oxidation numbers are based on assigning shared electrons to the more electronegative bonded atoms whereas formal charges are based on assigning shared electrons equally to bonded atoms. Each approach provides numbers suitable for certain uses, but neither approach gives the best estimates of charges on atoms in molecules and polyatomic ions, because bonded electrons are usually shared unequally in such species.

Professor Leland Allen of Princeton University has developed a more reasonable means of estimating atom charges that allows for unequal sharing of bonded electrons. The atom charges that are obtained using Allen's method are fractional charges and are therefore called **partial charges**. *According to Allen,*

partial charge   = group number of A - number of lone pair
   on atom A   electrons on A - $(\chi_A/\Sigma\chi)$(number of bonding
                         electrons shared by A)   (9c)

where $\Sigma\chi$ is the sum of the electronegativities of atom A and the atom to which it is bonded.

• In the case of LiF, the partial charge of Li = 1 - 0 - $[\chi_{Li}/(\chi_{Li} + \chi_F)](2) = 1 - 0 - (1.0/5.0)(2) = 0.60+$.

• In the case of $Cl_2C=O$, the partial charge of C = 4 - 0 - $[\chi_C/(\chi_C + \chi_{Cl})](2) - [\chi_C/(\chi_C + \chi_{Cl})](2) - [\chi_C/(\chi_C + \chi_O)](4) = 4 - 0 - (2.5/5.5)(2) - (2.5/5.5)(2) - (2.5/6.0)(4) = 0.52+$.

In both of these cases, the sum of the partial charges for all the atoms must equal zero, because there is no overall charge

*for compounds. In the case of atoms in polyatomic ions, the sum
of the partial charges must equal the charge of the ion.*

Partial charges calculated using Allen's method are in agreement
with those calculated using sophisticated computer programs.
Furthermore, partial charges reflect the differing abilities of
atoms to attract shared electrons and can be used to successfully predict points at which molecules are likely to approach
one another during reactions.

9.5 Molecular Shape

The physical and chemical properties of compounds are related to
their structures. The **valence shell electron-pair repulsion
(VSEPR) model** can be used to successfully predict the shapes of
most molecules and polyatomic ions. *The VSEPR model states that
repulsions between the bond pairs and lone pairs in the valence
shell of an atom determine the angles between that atom and the
atoms bonded to it. This means the electron-pair geometry about
an atom is determined by the total number of regions of electron
density about that atom, the number of bonding regions plus the
number of nonbonding (lone pair) regions.*

*The electron-pair geometries that are commonly encountered when
working with molecules and ions composed of nonmetals are:*

Number of electron regions	Electron-pair geometry
2	linear
3	triangular planar
4	tetrahedral
5	triangular bipyramidal
6	octahedral

*The angles between atoms in these geometries when all of the
positions are occupied by atoms are given in Fig. 9.10 in the
text.* ✶ Try preparing a "flash card" summarizing this information and memorizing
it.

*When counting electron regions to use in predicting electron-
pair geometries, count multiple bonds as just one region,
because all the electrons of the multiple bond occupy the same
general region between the bonded atoms*. Thus, there are three
electron regions about the C atom in $Cl_2C=O$, and the electron-
pair geometry about the C atom is predicted to be triangular
planar.

Central Atoms with Bond Pairs and Lone Pairs

There are four electron regions about the N atom in $:NH_3$, and the electron-pair geometry about the N atom is predicted to be tetrahedral, because the electron-pair geometry includes the spatial arrangements of both bonding and nonbonding (lone pair) regions. However, the **molecular geometry** of a molecule or ion includes just the arrangements of atoms in space **and therefore differs from the electron-pair geometry whenever there are lone pairs of electrons about the central atom.** The molecular geometry of $:NH_3$ is triangular pyramidal, because the N atom is at the apex of a pyramid that has three H atoms for a triangular base. Furthermore, the H-N-H bond angle (107.5°) is less than the bond angle that is predicted for a tetrahedral overall electron-pair geometry (109.5°). **This is because lone pairs of electrons occupy larger volumes than bond pairs and cause greater repulsions than bond pairs, so the bonds collapse until the lone pair-bond pair repulsions are equal to the bond pair-bond pair repulsions. The relative order of repulsions is always lone pair-lone pair > lone pair-bond pair > bond pair-bond pair within a given molecule or ion.**

The common molecular shapes for molecules and ions are shown in Figs. 9.12 and 9.13 in the text. **These shapes can be seen by covering up the positions of the lone pairs in the overall electron-pair geometries and focussing on just the positions of the atoms. They are:**

Total number of electron regions	Bond regions	Lone pair regions	Molecular shape
3	3	0	triangular planar
	2	1	angular
4	4	0	tetrahedral
	3	1	triangular pyramidal
	2	2	angular
5	5	0	triangular bipyramidal
	4	1	seesaw
	3	2	T shaped
	2	3	linear
6	6	0	octahedral
	5	1	square pyramidal
	4	2	square planar

∗ Try preparing a "flash card" for each total number of electron regions and memorizing this information.

Notice that *lone pairs always occupy the planar positions in triangular bipyramidal electron-pair arrangements. This is due to the fact that electron pairs in these positions interact with just two other electron pairs at 90° and therefore experience less repulsion than they would in the up-and-down axial positions that interact with three other electron pairs at 90°. Similarly, the placement of lone pairs opposite each other in the case of four bond regions and two nonbonding regions leads to less repulsion in an octahedral arrangement than would be encountered with the lone pairs adjacent to one another.*

9.6   Molecular Polarity

Molecules which are not totally symmetrical are polar molecules. In a **polar molecule**, electron density accumulates toward one side of the molecule, giving that side a slight negative charge, δ_-, and the other side a slight positive charge of equal value, δ_+. The extent to which polar molecules align themselves with the charges on a pair of oppositely charged plates depends on their **dipole moment**, μ, which is defined as the product of the magnitude of these partial charges and the distance by which they are separated and expressed in units called debyes (D). *The more polar the molecule, the higher its dipole moment. Furthermore, the more polar the molecule, the stronger the forces of attraction between the positive end of one molecule and the negative end of another molecule, and these types of interactions can determine whether a compound is a solid, liquid or gas at room temperature and affect other properties as well.*

To predict whether a molecule having a central atom C and terminal atoms T, CT_n, will be polar or nonpolar:

- draw an acceptable Lewis structure

- predict the electron-pair geometry

- determine whether the molecule is totally symmetrical

   • If all of the terminal atoms T are the same and there are no lone pairs of electrons around the central atom, the molecule is totally symmetrical and nonpolar.

   • If there are different terminal atoms or lone pairs of electrons that are opposite each other, the mole-

cule still meets the criterion of being totally
symmetrical and is nonpolar.

• If the molecule is not totally symmetrical, it is
polar. (Any molecule having just one lone pair of
electrons about the central atom necessarily belongs
in this category.)

9.7 Key Expressions

Enthalpy of Gas Phase Reaction - Bond Energy Expression:

ΔH°_{rxn} = ΣD(bonds broken) - ΣD(bonds formed)

Atom Formal Charge Expression:

atom formal = group number - number of lone pair electrons
charge   - ½(number of bonding electrons)

Atom Partial Charge Expression:

partial charge = group number of A - number of lone pair
on atom A   electrons on A - $(\chi_A/\Sigma\chi)$(number of bond
                 ing electrons shared by A)

9.8 Hints for Answering Questions and Solving Problems

Predicting Molecular Shapes:

The following steps can be used to predict the molecular
shapes of molecules and ions:

• Draw an acceptable Lewis structure.

• Count the number of bond regions and lone pair
regions about the central atom, and use the information given on p. 118 to predict the molecular geometry
about the central atom.

Predicting Molecular Polarities:

The following steps can be used to predict the polarities
of molecules:

• Draw an acceptable Lewis structure.

• Count the number of electron regions about the central atom, and use the information given on p. 117 to predict the electron-pair geometry about the central atom.

• Place the bonded atoms and lone pairs in their proper positions in the predicted electron-pair geometry.

• Determine whether the electron-pair arrangement is totally symmetrical. If it is, the molecule is nonpolar. If it is not, the molecule is polar.

Chapter 9

CHAPTER 10

BONDING and MOLECULAR STRUCTURE:
ORBITAL HYBRIDIZATION, MOLECULAR ORBITALS, and METALLIC BONDING

Lewis structures show the number of bonding and nonbonding pairs of electrons in molecules and can be used to rationalize the properties of covalent bonds and predict molecular shapes and polarities. However, electrons reside in orbitals, and this chapter describes the valence bond and molecular orbital theories of how atoms use orbitals to form covalent bonds. The merits of these approaches are discussed on p. 458 in the text.

10.1 Valence Bond Theory

According to **valence bond theory**, covalent bonding occurs when there is **orbital overlap** between two atoms and a maximum of two electrons of opposite spin occupy the overlapping orbitals. *The orbital overlap is caused by the attractive forces that exist between the nuclear charge of one of the bonded atoms and the electron cloud of the other and vice versa.* This mutual attraction leads to a decrease in potential energy in accord with Coulomb's law (p. 119 in text). However, at distances closer than the equilibrium internuclear distance of 74 pm in H_2, nuclear charge-nuclear charge and electron-electron repulsions begin to outweigh the attractions and the potential energy reaches a minimum, as shown in Fig. 9.2 in the text..

The overlap of two s orbitals, one from each atom, leads to a **sigma (σ) bond** which has its greatest electron density along the bond axis. *Sigma bonds are also formed by the overlap of an s orbital with a p orbital and by the end-to-end overlap of two p orbitals, as shown in Fig. 10.2 in the text.*

The orbitals that were used to write the electron configurations of isolated atoms can be used to describe the bonding in diatomic molecules, such as H_2, F_2 and HF. In the case of HF, overlap between the 1s orbital of H and the 2p orbital of F that contains just one valence electron yields a σ bond in which the shared pair of electrons is under the influence of the nuclei of both atoms. *However, we have difficulty accounting for the number of bonds formed by atoms and the shapes of most molecules using these orbitals that were adequate for isolated atoms. The alternatives are to describe chemical bonding in terms of hybrid atomic orbitals or molecular orbitals.*

Hybrid Orbitals

Linus Pauling suggested the approach of other atoms causes the valence orbitals of the central atom to combine to give **hybrid atomic orbitals** of new energy and direction in space. The driving force for **orbital hybridization** is presumably to **1)** provide orbitals of new shape that give more effective overlaps and therefore form stronger bonds and **2)** provide orbitals at larger angles that minimize the repulsions among the electron pairs in the valence shell of the central atom in accord with the precepts of VSEPR. For example, *the formation of sp^3 (s-p-three) hybrid orbitals for C accounts for the formation of four bonds to H atoms and the VSEPR predicted and experimentally observed H-C-H bond angle of 109.5° for CH_4.* The bonding in CH_4 is discussed in detail on pp. 461 and 462 in the text.

The mathematical combination of an s and p orbital from the same atom yields two hybrid atomic orbitals which are labelled sp and are located 180° apart along the axis of the p orbital combined. The mathematical combination of an s orbital and two p orbitals from the same atom yields three hybrid sp^2 (s-p-two) atomic orbitals 120° apart in the plane that is formed by the intersection of the axes of the two p orbitals involved. For example, if the p_x and p_y orbitals are combined, the three sp^2 hybrids would be located in the xy plane and be located 120° apart, *since this is the angle that provides the greatest separation and minimum repulsion between electrons in orbitals.*

The number of hybrid atomic orbitals formed is always equal to the number of atomic orbitals used in the combination. Furthermore, superscripts are used to indicate just how many atomic orbitals of each type are combined. This is why the combination of one s orbital and one p orbitals yields two hybrid that are labelled as sp hybrids and the combination of one s orbital and two p orbitals yields three hybrid atomic orbitals that are labelled as sp^2 hybrids. It is important to note that **sp hybridization leaves two p orbitals unchanged and sp^2 hybridization leaves one p orbital unchanged. These unchanged p orbitals can be used to form additional bonds called π bonds.**

Hybrid Orbitals Involving d Orbitals

Elements from the third and higher periods can accommodate more than an octet of electrons in their valence shells. However, there are only four atomic orbitals of the s and p type in the valence shell of these atoms, and the extra orbitals that are required to exceed an octet must come from the d orbitals of the same shell. Sulfur, for example, has its valence electrons in 3s and 3p orbitals but readily uses 3d orbitals to form $:SF_4$ and SF_6. *The mathematical combination of one s orbital, three p orbitals and one d orbital yields five sp^3d hybrid atomic* *orbitals pointing towards the corners of a triangular bipyramid. Similarly, the mathematical combination of one s orbital, three p orbitals and two d orbitals yields six sp^3d^2 hybrid orbitals pointing towards the corners of an octahedron.*

The hybrid orbital sets that are commonly postulated for covalent compounds of the nonmetals are summarized in Table 10.1 in the text and in a table on p. 475 in the text. ***The important characteristics of these hybrid orbital sets are:***

Orbitals combined	Type of hybrids formed	Number of hybrids formed	Geometry of hybrids formed
s+p	sp	2	linear
s+p+p	sp^2	3	triangular planar
s+p+p+p	sp^3	4	tetrahedral
s+p+p+p+d	sp^3d	5	triangular bipyramid
s+p+p+p+d+d	sp^3d^2	6	octahedral

The shapes of these respective hybrids are shown in Fig. 10.10 in the text.

Multiple Bonding

A double bond consists of a σ and a π bond, and a triple bond consists of a σ and two π bonds. **Pi (π) bonds** arise from the sideways overlap of *p* atomic orbitals, meaning the overlap region is above and below the internuclear axis, and the π-bonding electrons are above and below the bond axis. *The formation of p-p π bonds requires the use of unchanged p orbitals. This means that if the central atom in a multiply bonded Lewis structure is from the second period, the atoms involved in the multiple bonds must be sp or sp^2 hybridized.*

The bonding in the two classic examples of molecules with double and triple bonds, C_2H_4 and C_2H_2, is described in the text, and the orbital overlaps for these molecules are pictured in Figs. 10.12-14 in the text. *Notice that hybrids are involved in the formation of each σ bond whereas π bonds are formed by unchanged p orbitals.*

A Consequence of Multiple Bonding: Isomers

Atoms are free to rotate around a single, σ bond in any mole-cule. However, free rotation is not possible around multiple bonds under ordinary conditions, because this would require breaking the π bonds. A consequence of this is the existence of cis and trans isomers for compounds like XYC=CXY. The geometry about each C atom is triangular planar, and the X atoms can either be adjacent to each other, as in the *cis* form, or opposite each other, as in the *trans* form; see p. 474 in the text. **The cis and trans forms of compounds are called isomers and are examples of geometric isomers or more generally stereoisomers.** In general, **isomers** are compounds that have the same formula but different structures.

10.2 Molecular Orbital Theory

Molecular orbital (MO) theory is an alternative to the valence bond theory of describing covalent bonding in terms of localized atomic orbitals or localized hybrid atomic orbitals. MO theory assumes the atomic orbitals of atoms in the molecule combine to form **molecular orbitals** that are spread out, or delocalized, over several atoms or even over the entire molecule. MO theory can account for the paramagnetism of O_2 whereas valence bond theory cannot.

Principles of Molecular Orbital Theory

In MO theory we begin with the given arrangement of atoms at known bond distances and determine the sets of molecular orbitals that can form from combining all the available orbitals of all the constituent atoms. The **first principle of molecular orbital theory** states that the number of molecular orbitals formed is always equal to the number of atomic orbitals brought by the atoms that have combined. *A number of electrons equal to the total number of valence electrons for all the atoms in the molecule is then assigned to these orbitals following the Pauli exclusion principle and Hund's rule.*

Bonding and Antibonding Orbitals in H_2:

Two molecular orbitals result from the addition and subtraction of the wave functions for the overlapping 1s orbitals in H_2. The addition combination leads to an increased probability of finding the electrons in the overlapping region, and the resulting sigma molecular orbital is called a **bonding molecular orbital** that is labelled the σ_{1s} to indicate its origin. The subtracting combination reduces the electron probability in the overlapping region and increases it on the outside of the molecule. It is an **antibonding molecular orbital** that has no counterpart in valence bond theory and is labelled the σ_{1s}^*.

The **second principle of molecular orbital theory** states that the bonding molecular orbital is lower in energy than the orbitals from which it is formed, and the antibonding orbital is higher in energy, as shown in Fig. 10.17 in the text. The **third principle of molecular orbital theory** states that the valence electrons of the molecule are assigned to orbitals of successively higher energy following the Pauli exclusion principle and Hund's rule. **Thus, the two electrons of H_2 are assigned to the σ_{1s} orbital with their spins paired, and the electron configuration of H_2 is written as $(\sigma_{1s})^2$.**

Bond Order:

The bond order in MO theory is calculated using:

Bond order = ½(number of electrons in bonding MOs − number of electrons in antibonding MOs)   (10a)

The only requirement for bonding with MO theory is to have an excess of electrons in bonding MOs compared to antibonding MOs. There is no requirement for electron pairs or octets, and fractional bond orders are allowed. The higher the bond order, the stronger the bond, as with valence bond theory.

Molecular Orbitals for Li_2 and Be_2:

A **fourth principle of molecular orbital theory** states that atomic orbitals combine to form molecular orbitals most effectively when the atomic orbitals are of similar energy. **This means 1s-2s combinations are not important for diatomic molecules of the second period elements, and only 1s-1s and 2s-2s combinations need to be considered for Li_2 and Be_2. The 1s core electrons of these elements are assigned to σ_{1s} and σ_{1s}^* orbitals and therefore offset one another. Only the valence electrons in the 2s and 2p orbitals of the atoms affect the net bonding in the diatomic molecules of the second period elements.**

Molecular Orbitals for Homonuclear Diatomic Molecules of Second Period Elements Having Valence Electrons in p Atomic Orbitals:

Molecules formed from two identical atoms, such as H_2 and O_2, are called **homonuclear diatomic molecules.**

Molecular Orbitals from Atomic p Orbitals:

The following atomic orbital combinations are possible for homonuclear diatomic molecules of elements of the second period having valence electrons in s and p atomic orbitals:

• s-s combinations to form sigma bonding and antibonding, σ_{2s} and σ_{2s}^*, molecular orbitals.

• end-to-end p-p combinations to form sigma bonding and antibonding, σ_{2p} and σ_{2p}^*, molecular orbitals.

• side-to-side p-p combinations to form two pi bonding and

two pi antibonding, π_{2p} and π_{2p}^*, molecular orbitals at right angles to one another.

These combinations are illustrated in Figs. 10.20 and 10.21 in the text, and the relative energies of the resultant molecular orbitals are shown in Fig. 10.22 in the text. ***Notice that the general order of energy is:***

$$\sigma_{2s} < \sigma_{2s}^* < \pi_{2p} = \pi_{2p} < \sigma_{2p} < \pi_{2p}^* = \pi_{2p}^* < \sigma_{2p}^*$$

The rationale for the positions of these molecular orbitals is given in "A Closer Look: Molecular Orbitals for Compounds Having *p*-Block Elements" on p. 483 in the text. However, these details are not important to being able to use the general order of energy to rationalize the properties of the homonuclear diatomic molecules of these elements.

Electron Configurations for Some Homonuclear Diatomic Molecules:

The general order of energy given above can be used to write the molecular orbital electron configurations for the homonuclear diatomic molecules of the second period elements and rationalize such properties as bond order, bond length, bond dissociation energy and magnetism. See Table 10.2 in the text, and notice that ***the σ_{1s} and σ_{1s}^* molecular orbitals formed by the 1s core electrons are not included, because they offset one another and do not contribute to the bonding in these molecules***. Also notice the ability of MO theory to account for the paramagnetism of O_2 whereas valence bond theory could not.

Resonance and π Bonds

According to valence bond theory, it was necessary to portray certain molecules and ions as composites of various resonance forms, because it was not possible to draw Lewis structures corresponding to their actual structures. However, MO theory is able to describe the actual structures in terms of molecular orbitals that are spread over the entire structures. In the case of ozone, O_3, there is a sigma bond between each terminal O atom and the central atom and a pi bond that is delocalized over all three atoms and naturally leads to a bond order of 1½ for each bond in accord with experimental facts. In the case of $CO_3{}^{2-}$, there is a filled pi molecular orbital that is delocalized over the entire ion that leads to a bond order of 1⅓, and in the case of C_6H_6, there are three filled pi molecular orbitals that lead to a bond order of 1½ for each carbon-carbon bond. The bonding in benzene, C_6H_6, is described in detail in "A Closer Look" on p. 510 in the text.

10.3 Metals and Semiconductors

The molecular orbital theory that is used to describe diatomic molecules can be extended to describe the properties of metals and semiconductors.

Conductors, Insulators and Band Theory

In a metal there are so many atoms close together that the molecular orbitals form a band of molecular orbitals having closely spaced energy levels; see Fig. 10.24 in the text. **The band is composed of as many energy levels as there are contributing atomic orbitals, and each can contain two electrons of opposite spin.** The band is spread over the atoms of the piece of metal accounting for the bonding in metals, and this theory of bonding is called the **band theory of bonding.**

The band of energy levels in a metal is only partly filled, and the highest filled level is called the **Fermi level**. *Small inputs of energy, such as thermal energy, can cause electrons to be promoted from the filled portion of the band to the unfilled portion causing the formation of singly occupied levels in both portions. Movement of electrons in these singly occupied levels in the presence of an applied electric field is responsible for the electrical conductivity of metals. Furthermore, the absorption of light and concurrent promotion of electrons is followed by emission of a photon of the same wavelength as the electrons return to their original energy levels, and this is why polished metal surfaces are reflective and appear lustrous.*

The highest occupied band of a substance is called its **valence band**. The valence band of a metal is only partly filled. However, the valence band of an **insulator** is completely filled and is separated from a band that is formed by its antibonding molecular orbitals and is called its **conduction band**. The energy separation between the valence and conduction bands of an insulator is called the **band gap** and is sufficiently great that electron promotion is not possible, and the solid does not conduct electricity.

Semiconductors

The band gap between the valence and conduction bands of **semiconductors** is less than the band gap obtained for insulators. *At least a few electrons can be promoted by modest inputs of energy, and electrical conduction can occur.* It is interesting to note that carbon as diamond is an insulator whereas silicon and germanium have the diamond structure and are semiconductors.

Pure silicon is an example of an **intrinsic semiconductor.** *The promotion of an electron to the conduction band of an intrinsic semiconductor creates a positive hole in the valence band. The semiconductor carries electrical charges because the electrons in the conduction band migrate in one direction and the positive holes in the valence band migrate in the opposite direction. A positive hole "moves" as an electron from an adjacent level falls into the hole and thus creates a new positive hole.*

The number of electrons in the conduction band of an intrinsic semiconductor depends on the magnitude of the band gap. The smaller the gap, the greater the number of electrons at a given temperature. The higher the temperature, the greater the number of electrons, as well.

Extrinsic semiconductors are materials whose conductivity is controlled by adding tiny numbers of atoms of different kinds as impurities called **dopants.** The addition of an atom with less valence electrons (IIIA element) to an intrinsic semiconductor (IVA element) creates an **acceptor level** just above the valence band, and electrons can be readily promoted from the valence band to the conduction band. *This leaves positive holes in the valence band that can move under the influence of an electric potential,* so the doped semiconductor is called a **p-type semiconductor.** On the other hand, addition of an atom with more valence electrons (VA element) creates a **donor level** just below the conduction band. *Promotion of these electrons to the conduction band creates a negative charge carrier,* so the doped semiconductor is called a **n-type semiconductor.**

10.4   Key Expressions

Molecular Orbital Theory Bond Order Expression:

Bond order = ½(number of electrons in bonding MOs - number of electrons in antibonding MOs)

10.5   Hints for Answering Questions and Solving Problems

Predicting Hybrid Orbitals Used by Central Atoms:

The orientations of the hybrid orbitals correspond to the electron-pair geometries that can be predicted by using the principles of the VSEPR theory that was considered in Chapter 9. Thus, the following steps can be used to predict the type of hybrids being used by central atoms:

- Draw an acceptable Lewis structure.

- Count the number of electron regions about the central atom of interest, and predict the electron-pair geometry about the central atom using:

Number of electron regions	Electron-pair geometry
2	linear
3	triangular planar
4	tetrahedral
5	triangular bipyramidal
6	octahedral

- Select the type of hybrids corresponding to the predicted electron-pair geometry using:

Type of hybrids	Number of hybrids	Geometry of hybrids
sp	2	linear
sp^2	3	triangular planar
sp^3	4	tetrahedral
sp^3d	5	triangular bipyramid
sp^3d^2	6	octahedral

Writing Electron Configurations for Diatomic Molecules of Second Period Elements:

The electron configurations for diatomic molecules of second period elements can be written by:

- counting the total number of valence electrons residing in the $2s$ and $2p$ orbitals of both atoms and

- placing a maximum of two electrons of opposite spins into the appropriate molecular orbitals using the general order of energy:

$$\sigma_{2s} < \sigma_{2s}^* < \pi_{2p} = \pi_{2p} < \sigma_{2p} < \pi_{2p}^* = \pi_{2p}^* < \sigma_{2p}^*$$

and taking care to place one electron in each of the energetically equivalent π_{2p} and π_{2p}^* orbitals before placing two electrons in either in accord with Hund's rule.

Chapter 10

CHAPTER 11

BONDING and MOLECULAR STRUCTURE: ORGANIC CHEMISTRY

The class of covalently bonded compounds that is based on carbon is called **organic compounds.** *Carbon itself exists as graphite, diamonds and buckyballs* (Fig. 3.5 in the text.). Graphite consists of layers of carbon atoms that are arranged in hexagonal rings within layers. Each carbon atom is therefore surrounded by a triangular planar arrangement of other carbon atoms and is thought to be sp^2 hybridized. This means there is delocalized π bonding within layers, and this is why graphite is able to conduct electricity. There is, however, no covalent bonding between layers. There are only weak forces of attraction between layers, so the layers are able to slide past each other with ease and graphite is a good lubricant. On the other hand, each carbon atom in diamond is surrounded by four other carbon atoms and is thought to be sp^3 hybridized. The structure is so rigid that diamond is one of the hardest substances known, and there are no delocalized electrons, so diamond is a not a conductor of electricity. It is an insulator, as was discussed in 10.3.

The simplest class of organic compounds is the carbon-hydrogen series of compounds called **hydrocarbons.** *These can be further classified as alkanes, alkenes, alkynes and aromatics based on the types of chemical bonds found in them.* This chapter describes the structure, bonding and chemical reactivity of hydrocarbons, alcohols, aldehydes, ketones, carboxylic acids, esters, amines and amides, as well as the chemistry of polymers.

11.1 Alkanes

The simplest alkane is methane, CH_4. Replacing one of the H atoms of methane with a $-CH_3$ group gives ethane, CH_3CH_3 or C_2H_6. Continuing to replace an end H atom with a $-CH_3$ group gives the series of **"straight-chain" alkanes** with unbranched carbon atoms that is listed in Table 11.2 in the text. *The general formula for alkanes is C_nH_{2n+2}. The systematic names of alkanes are based on the number of carbon atoms in the longest carbon atom chain*, as described in the "Closer Look" on p. 501 in the text.

Physical and Chemical Properties of the Alkanes

The alkanes are nonpolar, as are most hydrocarbons. The four lightest alkanes (methane, ethane, propane and butane) are gases at room temperature and pressure. Those having longer unbranched carbon chains and higher molar masses are liquids or solids.

Alkanes readily burn in air to give carbon dioxide and water in highly exothermic reactions but fail to do so at room temperature and pressure. This is because the rate of reaction is too low. After studying Chapter 15, you will be able to understand why this is so.

Under the proper set of conditions, the H atoms of alkanes can be replaced by other atoms, such as halogen atoms. Clorofluorocarbons are used as refrigerants but appear to be involved in the depletion of the ozone layer that protects us from some of the ultraviolet radiation from the sun and are being replaced.

Bonding and Structure in Alkanes

Every carbon atom in an alkane is surrounded by four other atoms and is thought to be sp^3 hybridized. This results in zig-zag, or nonlinear, chains of carbon atoms in alkanes, as shown for butane and pentane on pp. 499 and 500 in the text.

Alkanes containing four or more carbon atoms can exist as structural isomers. Structural isomers are compounds that have the same formula but are structurally different because their atoms are connected differently. This is shown in the structures of the three isomers of C_5H_{12} on p. 500 in the text. As the number of carbon atoms in alkanes increases, the number of possible structural isomers increases rapidly.

Cycloalkanes

Alkanes can also exist in rings of sp^3 hybridized C atoms. These alkanes have the general formula C_nH_{2n} and are called **cycloalkanes. Rotation around carbon-carbon bonds is also possible in cycloalkanes, but it takes the form of bending the entire molecule.** The two possible forms that result are usually referred to as the "*chair*" and "*boat*" **conformations** of the molecule, because they resemble these shapes. These conformations of cyclohexane, C_6H_{12}, are shown on p. 501 in the text.

11.2 Alkenes and Alkynes

Hydrocarbons containing double bonds and having the general formula C_nH_{2n} are called **alkenes**, and those containing triple

---

bonds and having the general formula C_nH_{2n-2} are called **alkynes**. Alkenes and alkynes are said to be **unsaturated**, because the pi electrons of the multiple bonds can be used to form sigma bonds to attach additional atoms. These compounds are therefore widely used in industry to synthesize new compounds.

Naming Alkenes and Alkynes

The systematic names of the alkenes and alkynes are derived from the names of the corresponding alkanes by dropping the "ane" ending and adding "ene" for alkenes or "yne" for alkynes. The systematic names for the simplest alkene, C_2H_4, and the simplest alkyne, C_2H_2, are ethene and ethyne, but these compounds are almost always called their common names of ethylene and acetylene.

When there is more than one position possible for the position of the double or triple bond, the actual position is included in the name. To determine the position to be used in naming, we begin counting at the end of the carbon chain closest to the multiple bond. Thus, CH_3-CH_2-CH=CH-CH_3 is 2-pentene and not 3-pentene, because the double bond begins with the second carbon atom from the right.

Structure, Bonding and Isomerism

In alkenes, the substituents are arranged around the doubly bonded carbon atoms as planar triangles. In alkynes, the R-C≡C-R atom arrangement is linear. Valence bond theory accounts for this by postulating that the carbon atoms involved in the double bond in an alkene are sp^2 hybridized, and that the π bond is formed by the overlap of unhybridized p atomic orbitals, one from each C atom. Similarly, the carbon atoms involved in the triple bond in an alkyne are believed to be sp hybridized, and the two π bonds are believed to be formed from two unhybridized p atomic orbitals on each atom.

An important consequence of double bonding in alkenes is stereoisomerism. In **stereoisomers**, the atom-to-atom connections are the same, but the atoms are arranged differently in space. Stereoisomers containing two substituents of one kind and two of another kind are called **cis-isomers** when the like substituents are on the same side of the doubly bonded C atoms and **trans-isomers** when the like substituents are on opposite sides of the doubly bonded C atoms. The structures of *cis-* and *trans*-2-butene are shown on p. 504 in the text. Notice how these isomers have differing polarities and therefore differing melting and boiling points. *Cis-trans stereoisomers commonly have differing physical properties*.

Preparation of Alkenes and Alkynes

Ethylene is prepared commercially by using steam to "crack" the hydrocarbons found in natural gas and petroleum.

$$C_2H_6(g) \ \longrightarrow\text{steam}\longrightarrow \ C_2H_4(g) + H_2(g) \tag{11a}$$

More than 18 billion kilograms of ethylene are produced annually in the United States and about half of this amount is used to prepare polyethylene plastic. Ethylene is one of the top five chemicals produced in the United States, and the development of methods of producing relatively inexpensive ethylene enabled chemists to replace acetylene with ethylene for commercial processes.

Acetylene is prepared

from limestone:

$$CaCO_3(s) \rightarrow CaO(s) + CO_2(g) \tag{11b}$$

$$CaO(s) + 3 \ C(s) \rightarrow CaC_2(s) + CO(g)$$

$$CaC_2(s) + 2 \ H_2O(\ell) \rightarrow C_2H_2(g) + Ca(OH)_2(s)$$

and

from methane:

$$2 \ CH_4(g) \rightarrow C_2H_2(g) + 3 \ H_2(g) \tag{11c}$$

but both processes involve high temperatures and are costly.

Addition Reactions

One of the most common reactions of alkenes (and alkynes) is addition reactions in which reagents add to the carbon atoms of the C=C double bond. The general equation for addition reactions of ethylene is:

$$C_2H_4 + XY \rightarrow XCH_2CH_2Y \tag{11d}$$

where XY can be H_2, H_2O, X_2 or HX and X is a halogen. When the reagent is H_2, the process is called **hydrogenation** and the product is an alkane. ✱ Try preparing a "flash card" illustrating the general equation for addition reactions of ethylene and memorizing it.

11.3 Aromatic Compounds

Benzene, C_6H_6, toluene, $C_6H_5CH_3$, and napthalene, $C_{10}H_{10}$, are members of a class of compounds called **aromatics**, because they usually have a pleasant odor. *The structures of these compounds, shown on p. 508 in the text, suggest they are unsaturated compounds, like the alkenes. However, reactions of benzene with the halogens give substitution products rather than addition products. That is, the halogen replaces one H atom from benzene and does not add to the carbon-carbon double bond of benzene. This observation lead August Kekulé to suggest in 1872 that benzene exists as a planar resonance hybrid of two structures and therefore has equivalent carbon-carbon and carbon-hydrogen bonds;* see p. 509 in the text.

Modern experimental methods have indeed shown that the each carbon-carbon bond length in benzene is equivalent and intermediate between carbon-carbon single and double bond lengths. Thus, chemists usually place a circle in the middle of a hexagonal ring to abbreviate the structure of benzene and indicate that the pi-bonding electrons are delocalized over the six carbon atoms of the ring. The molecular theory method of describing the delocalized π bonding in benzene is discussed in "A Closer Look" on p. 510 in the text.

Naming Aromatic Compounds

Most aromatic compounds are named as derivatives of benzene. When a benzene ring bears two substituents, the substituents are named in alphabetical order and the positions of the substituents are indicated by calling the position of the first substituent carbon atom 1 and using the lowest possible number to indicate the position of the second substituent relative to the first. Thus, the dichlorobenzene having chlorine atoms on adjacent carbon atoms is named 1,2 dichlorobenzene rather than 1,6 dichlorobenzene. *The terms ortho, meta, and para are also used to represent the 1,2; 1,3 and 1,4 positions,* so 1,2 dichlorobenzene is also called *o*-dichlorobenzene.

11.4 Alcohols

If one or more of the H atoms of an alkane or alkene is replaced by an OH (hydroxyl) group, the result is an alcohol, ROH. Methanol, CH_3OH, ethanol, C_2H_5OH, and ethylene glycol, $HOCH_2CH_2OH$, are examples of commercially important alcohols.

Naming Alcohols

Systematic names of alcohols are derived from the names of the corresponding alkanes by dropping the "e" ending and adding "ol." When necessary, a numerical prefix is used to designate the position of the OH group; see Table 11.2 in the text.

Alcohols can be classified into three categories based on the number of carbon atoms bonded to the C atom bearing the OH group:

  • **Primary alcohols** have one carbon atom and two hydrogen atoms attached to the C atom bearing the OH group.

  • **Secondary alcohols** have two carbon atoms and one hydrogen atom attached to the C atom bearing the OH group.

  • **Tertiary alcohols** have three carbon atoms and no hydrogen atoms attached to the C atom bearing the OH group.

✷ Try preparing a "flash card" showing the connectivities of the carbon atoms bearing OH groups in these classes of alcohols and memorizing it.

Some Chemistry of Alcohols

Alcohols have the general formula ROH, where R is some organic group, and therefore bear some resemblance to water, HOH. Thus, alcohols react with active metals analogous to water.

$$ROH(\ell) + Na(s) \rightarrow \text{½ } H_2(g) + NaOR(aq) \tag{11e}$$

$$HOH(\ell) + Na(s) \rightarrow \text{½ } H_2(g) + NaOH(aq) \tag{11f}$$

On the other hand, alcohols react with hydrogen halides to produce organic halides whereas water reacts with hydrogen halides to produce acid solutions.

$$ROH(\ell) + HBr(g) \rightarrow RBr(aq) + H_2O(\ell) \tag{11g}$$

$$HOH(\ell) + HBr(g) \rightarrow H_3O^+(aq) + Br^-(aq) \tag{11h}$$

Concentrated sulfuric acid can react with alcohols to remove H and OH from adjacent carbon atoms to produce the corresponding alkene. Under other conditions, concentrated sulfuric removes the H atom from one alcohol molecule and the OH group from another alcohol molecule to form an **ether** and water.

$$2 \text{ ROH} \rightarrow R-O-R + H_2O \tag{11i}$$

An alcohol has an R group in place of one H atom of water, and an ether has two R groups, which can differ, in place of both H atoms of water.

11.5 Compounds With A Carbonyl Group

A double-bonded carbon-oxygen, C=O, grouping is called a **carbonyl group**.

   • **Aldehydes** have one organic group and one H atom attached to the carbon atom of a carbonyl group.

   • **Carboxylic acids** have one organic group and one OH group attached to the carbon atom of a carbonyl group.

   • **Ketones** have two organic groups attached to the carbon atom of a carbonyl group.

✷ Try preparing a "flash card" summarizing this information and memorizing it.

The oxidation of a primary alcohol produces an aldehyde which produces a carboxylic acid with further oxidation. The oxidation of a secondary alcohol produces a ketone.

Aldehydes and Ketones

Aldehydes and ketones have pleasant odors and are often used as the basis of fragrances. *Systematic names of aldehydes are obtained by dropping the ending "e" from the name of the corresponding alkane and adding "al" whereas systematic names of ketones are obtained by dropping the ending "e" from the name of the corresponding alkane and adding "one."* However, the alde-hydes and ketones listed in Table 11.7 in the text are usually called their common names.

Aldehydes and ketones are the oxidation products of primary and secondary alcohols, respectively. Thus, the reduction products of aldehydes and ketones are primary and secondary alcohols, respectively. Common reducing agents, such as $NaBH_4$ and $LiAlH_4$, can be used for these reductions.

Carboxylic Acids

Carboxylic acids are the end products of oxidation of primary alcohols. However, acetic acid, the most important carboxylic acid commercially, is currently being produced by reacting methanol and carbon monoxide in the presence of a catalyst:

$$CH_3OH(\ell) + CO(g) \xrightarrow{\text{catalyst}} CH_3COOH(\ell) \qquad (11j)$$

Only the H atom that is attached to the O atom in acetic acid is sufficiently positively charged due to electronegativity differences to act as an acidic H atom, a proton. Thus, acetic acid is a monoprotic acid.

The systematic names of carboxylic acids are obtained by dropping the ending "e" from the name of the corresponding alkane and adding "oic" and the word "acid." However, carboxylic acids often have common names that are derived from the Latin or other name for their natural occurrences. Examples of names of carboxylic acids can be found in Table 11.8 in the text.

Esters

Carboxylic acids react with alcohols in the presence of strong acids to produce esters. An **ester** is produced when the OH group of the acid combines with the H atom of the OH group of the alcohol to form water and a bond is formed between the C atom of the carbonyl group of the acid and the O atom of OR remaining from the alcohol.

$$RCO\mathbf{OH} + \mathbf{H}OR' \xrightarrow{H^+} RCOOR' + \mathbf{HOH} \qquad (11k)$$

The strong acid furnishes H^+ ions which act as catalysts.

The two-part name of an ester is given by 1) the name of the alkyl group from the alcohol and 2) the name of the carboxylate group derived from the acid. Examples of names of esters can be found in Table 11.10 in the text.

The most important reactions of esters are their hydrolysis reactions with water. These reactions are generally conducted in the presence of a base like NaOH and yield the sodium salt of the carboxylic acid and the alcohol. The carboxylic acid can be recovered by reacting the sodium salt with strong acid, because carboxylic acids are weak acids and carboxylate anions are weak bases that are readily protonated.

11.6 Fats and Oils

All fats are triesters of long-chain "fatty acids" with glycerol, which is 1,2,3-propanetriol. The R groups from the fatty acids can be the same or different within the same fat molecule and can be saturated or unsaturated organic groups. If the group is unsaturated, it can be monounsaturated or polyunsaturated, depending on whether it contains one or more carbon-carbon double bonds.

In general, saturated fatty acids lead to fats (solids) and unsaturated fatty acids lead to oils (liquids). The reason for this is that saturated alkane chains are flexible and can pack close to one another resulting in solids whereas unsaturated chains contain double bonds which are not as flexible.

Fats in Food

Nutritionists recommend that we receive no more than 30% of our calories from fats and that no more than one-third of this be from saturated fat. However, fat makes up about 40% of the energy intake in the average American diet.

Soap

The hydrolysis of an ester is sometimes called a **saponification reaction** (from the Latin word *sapo*, meaning soap), because hydrolysis of a fat or oil gives a salt of a long-chain carboxylic acid, like sodium stearate, and glycerol. *Compounds, such as sodium stearate, that have a long-chain hydrocarbon portion that mixes with oil and greasy substances and an ionic end that mixes with water are soaps.* Interestingly, potassium salts tend to have lower melting points than their sodium counterparts because of the larger size of K^+ compared to Na^+ and are therefore used in liquid soaps.

11.7 Amines and Amides

Amines are derivatives of ammonia in which one or more of the H atoms of NH_3 have been replaced with organic R groups.

- **Primary amines** have the general formula RNH_2,

- **secondary amines** have the general formula R_2NH and

- **tertiary amines** have the general formula R_3N.

Amines are therefore bases and react with acids to form water-soluble ionic salts, like NH_3 ammonia reacts with HCl to form water-soluble NH_4Cl. ✶ Try preparing a "flash card" showing the general formulas of these classes of amines and memorizing it.

An ester is a derivative of a carboxylic acid and an alcohol. An **amide** is a derivative of a carboxylic acid and an amine.

$$RCOOH + HOR' \xrightarrow{-H^+} RCOOR' + HOH \qquad (11\ell)$$
$$RCOOH + HNR_2 \rightarrow RCONR_2 + HOH \qquad (11m)$$

The amide grouping is found in proteins, synthetic polymers, saccharin (p. 495 in the text) and N-acetyl-*p*-aminophenol, an analgesic that is known by the generic name acetaminophen and sold as Datril, Momentum, Tylenol, etc. Hence, it is a very important grouping.

11.8   Synthetic Organic Polymers

In 1976, synthetic polymers surpassed steel as the most widely used material in the United States. Thus, synthetic polymers are a vital part of our everyday experience, and chemists are still developing polymers with new and improved properties for our use.

The word "**polymer**" means "many parts" (from the Greek, *poly* and *meros*). Polymers are made by chemically joining many small molecules together to form **macromolecules** having molar masses that range from the thousands to the millions. The small molecules that are used to synthesize the polymers are called **monomers**.

Polymers can be classified according to their:

- *response to heat*,

    - **Thermoplastics** undergo reversible changes, such as softening and hardening, when heated and cooled.

    - **Thermosetting plastics** are plastic when first heated but then form solids which cannot be softened and reformed without extensive degradation when reheated.

- *end uses* as plastics, fibers, elastomers, coatings and adhesives, and

- *method of synthesis*.

    - **Addition polymers** are made by joining monomers directly.

    - **Condensation polymers** are made by joining monomers by splitting out small molecules, usually water, between them, as in the formation of esters and amides.

∗ Try preparing a "flash card" summarizing the characteristics of thermoplastics and thermosetting plastics and a second "flash card" summarizing the characteristics of addition and condensation polymers and memorizing them.

Addition Polymers

Addition polymerization, also called **chain growth polymerization**, is characterized by the fact that intermediates in the process are often **free radicals** containing unpaired electrons or other highly reactive species that cannot be isolated. Once polymerization is started, the chain grows rapidly.

Polyethylene:

The monomers for addition polymers normally contain one or more double bonds. The simplest monomer of this group is ethylene, C_2H_4, and polymerization begins with breaking the π bond of the carbon-carbon double bond to give an unpaired electron that is a reactive site at each end of the molecule. This is called the **initiation step** and is accomplished by using organic peroxides, RO-OR, that are unstable and break apart to give RO· free radicals. The free radicals resulting from the reaction between ethylene and the peroxide quickly combine to produce **polyethylene** which is composed of 1,000 to 50,000 monomer units. *The properties of the polyethylene formed vary with the temperature and pressure at which it is formed, so polyethylene is the world's most widely used polymer because of its wide range of properties.* See Table 11.13 in the text.

Polystyrene, Polyvinyl Chloride, and Other Addition Polymers:

If one of the H atoms of ethylene is replaced by another group, addition polymers can still be made, but they have different properties. Several examples are given in Table 11.13 in the text.

Natural and Synthetic Rubber:

Vulcanized rubber contains short chains of sulfur atoms that bond together the polymer chains of natural rubber. The sulfur chains help to align the polymer chains, so the material does not undergo a permanent change when stretched but returns to its original shape and size when the stress is removed. Substances that behave this way are called **elastomers**.

In the 1920's, German chemists discovered how to polymerize 1,3-butadiene to form the polybutadiene that is used in tires, hoses and belts. *In 1955, chemists at Goodyear and Firestone almost simultaneously discovered how to use special catalysts to prepare synthetic poly-cis-isoprene, which is structurally*

---

identical to natural rubber but not as cheap as natural rubber.

Copolymers

Copolymers are produced by polymerizing a mixture of two or more monomers. The most important synthetic rubber produced in the United States is produced using a 3-to-1 mole ratio of butadiene to styrene to make styrene-butadiene rubber (SBR), which is used to make tires and bubble gum.

Condensation Polymers

A chemical reaction in which two molecules react by splitting out or eliminating a small molecule is called a **condensation reaction.** *This type of polymerization reaction requires the presence of two different kinds of reactive functional groups on two different molecules.*

Polyesters:

A molecule with carboxylic acid groups at each end, such as terephthalic acid, and another molecule with alcohol groups at each end, such as ethylene glycol, can react with each other at both ends to form a **polyester**. The important uses of polyesters are described on p. 535 in the text.

Polyamides:

A molecule with carboxylic acid groups at each end and another molecule with amine groups at each end can react with each other at both ends to form a **polyamide**. The most familiar example of a polyamide is Nylon 6,6, which is prepared from adipic acid (a diacid) and hexamethylenediamine (a diamine). Kevlar, which is used to make bulletproof vests and fireproof garments, is an other familiar polyamide which is prepared from terephthalic acid and *p*-phenylenediamine.

New Polymer Materials:

Reinforced plastics, which contain fibers embedded in a matrix of a polymer, are called **composites.** *The use of a polymer matrix prevents the fibers from bending or buckling and results in a material that is stronger than steel.* Thus, reinforced plastics are used as body panels for the General Motors Saturn and Corvette automobiles, and there are numerous other uses of composites, such as the use of graphite-epoxy composites to make skis, tennis racquets, golf-club shafts and fishing rods.

11.9 Hints for Answering Questions and Solving Problems

Identifying Alkanes, Alkenes, Alkynes and Aromatics:

- Alkanes have only carbon-carbon single bonds and the general formula C_nH_{2n+2}.

    - Cylcoalkanes have only carbon-carbon single bonds and the general formula C_nH_{2n}.

- Alkenes have one or more carbon-carbon double bonds and the general formula C_nH_{2n}.

- Alkynes have one or more carbon-carbon triple bonds and the general formula C_nH_{2n-2}.

✳ Try preparing a "flash card" for each type of hydrocarbon that contains this information and memorizing them.

Identifying Alcohols, Aldehydes, Ketones, Carboxylic Acids, Esters, Amines and Amides:

- Alcohols contain the OH group and have the general formula ROH.

- Aldehydes contain the C=O group and have both an R and H attached to the C atom of the C=O group (RCHO).

- Ketones contain the C=O group and have two R groups attached to the C atom of the C=O group (RCOR).

- Carboxylic acids contain the C=O group and have one R group and one OH group attached to the C atom of the C=O group (RCOOH).

- Esters contain the C=O group and have one R group and one OR group attached to the C atom of the C=O group (RCOOR).

- Amines have one to three of the H atoms of ammonia, NH_3, replaced by R groups.

- Amides contain the C=O group and have one R and one NR_2 group attached to the C atom of the C=O group.

✳ Try preparing a "flash card" for each of these types of compounds that contains this information and memorizing them.

Chapter 11

CHAPTER 12

GASES

A qualitative introduction to the kinetic-molecular theory of matter was given in Section 1.1 in conjunction with an introduction to the states of matter. This chapter focusses on the gaseous state of matter and the laws which pertain to the behavior of gases. Gases behave remarkably similar, so these laws are applicable to virtually all gaseous substances.

12.1   The Properties of Gases

Four quantities are required to describe the state of a gas:

- the quantity of the gas, n, in moles,

- the temperature of the gas, T, in kelvins,

- the volume of the gas, V, in liters and

- the pressure of the gas, P, in atmospheres.

Gas Pressure

Pressure is force per unit area. The pressure of the atmosphere can be measured by using a crude **barometer**, which can be made by filling a tube with a liquid and inverting it into a dish containing the same liquid; see Fig. 12.1 in the text. Mercury is generally used because of its high density ($13.6 \ g/cm^3$). The pressure exerted by a column of mercury that is 760 mm Hg high exactly counterbalances the **standard atmospheric pressure** of 1 atmosphere (atm), so

$$1 \ atm = 760 \ mm \ Hg = 101.325 \times 10^3 \ Pa \qquad (12a)$$

where 1 Pa represents 1 **pascal**, the official SI unit of pressure. A unit called a **bar** is also used as a unit of pressure: 1 bar = 100,000 Pa, so 1 atm = 1.013 bar. The units pascal and bar are not widely used in chemistry, though thermodynamic data given in Chapters 6 and 20 and in Appendix K are given for gas pressures of 1 bar. The unit **torr** is often used interchangeably with the unit **mm Hg**, in honor of Torricelli, the inventor of the barometer in 1643: 1 torr = 1 mm Hg.

The use of barometers to measure atmospheric pressures and manometers and tire gauges to measure pressures of gases is described in "A Closer Look" on p. 549 in the text. Notice that *the pressure that is exerted by the column of liquid in a barometer is given by*:

$$P \propto \text{(height of column)} \times \text{(density of liquid)} \qquad \text{(12b)}$$

A column of water, density = 1.0 g/cm^3, would need to be 34 feet high to equal 1 atm. Also notice that a *tire gauge measures difference in pressures, since the atmosphere opposes the extension of the pressure scale.* Tire gauge pressure is given as

$$P_{gauge} = P_{\text{air in tire}} - P_{atmosphere} \qquad \text{(12c)}$$

and is expressed in pounds per square inch, **psi**: 14.7 lb/in^2 = 1 atm.

12.2 The Gas Laws: The Experimental Basis

Experimentation in the 17th and 18th centuries led to three gas laws that provide the basis for our understanding of gas behavior.

The Compressibility of Gases: Boyle's Law

The **compressibility** of gases refers to the fact that the volume of a given sample of gas at a fixed temperature can be de-creased. Robert Boyle studied the compressibility of gases in 1661 and observed that *the volume of a fixed amount of gas at a given temperature is inversely proportional to the pressure exerted on the gas.* Because all gases behave in this manner, this principle is known as Boyle's law.

The formal statement of **Boyle's law** is: *For a given quantity of gas at a given temperature, the product of pressure and volume is a constant.* This can be expressed mathematically as

$$P_1 V_1 = P_2 V_2 \qquad \text{(12d)}$$

for two different sets of pressure-volume conditions. Thus, *this form of Boyle's law is useful when we want to determine what happens to the volume of a given quantity of gas when the pressure changes at constant temperature or what happens to the pressure of a given quantity of gas when the volume changes at constant temperature.* ✳ Try preparing a "flash card" summarizing Boyle's observation, the formal statement of Boyle's law and the mathematical expression pertaining to two sets of pressure-volume conditions for a fixed quantity of gas at

---

constant temperature and memorizing it.

The Effect of Temperature on Gas Volume: Charles's Law

In 1787, Jacques Charles discovered *the volume of a given quantity of gas at constant pressure increased with increasing temperature. When plots of volume versus temperature are extended toward lower temperatures, they all extrapolate to zero volume at -273.15°C.* Thus, in 1884, Lord Kelvin suggested a temperature scale having -273.15°C as its zero point. This scale is called the **Kelvin scale**, and the units are called **kelvins**. The Kelvin scale is also known as the **absolute temperature scale**, because 0 K (-273.15°C) is believed to be the lowest temperature possible. Hence, 0 K is also referred to as **absolute zero**. *One kelvin is equivalent to one degree Celsius, so the only difference between the two scales is 273.15:* $T(K) = t(°C) + 273.15$ or simply $T(K) = t(°C) + 273$. <u>*Temperature is always expressed in kelvins when working with gases*</u>. ✳ Try preparing a "flash card" giving the formula for converting $t(°C)$ to $T(K)$ and memorizing it.

The formal statement of **Charles's law** is: *For a given quantity of gas at a given pressure, volume is directly proportional to its absolute temperature.* This means $V =$ constant x T, *so volume divided by absolute temperature is a constant and*

$$\frac{V_1}{T_1} = \frac{V_2}{T_2} \qquad (12e)$$

for two different sets of volume-absolute temperature conditions. *This form of Charles's Law is useful when we want to determine what happens to the volume of a given quantity of gas at constant pressure when the temperature is changed or what happens to the absolute temperature when the volume of a given quantity of gas at constant pressure is changed*. ✳ Try preparing a "flash card" summarizing Charles's observation, the formal statement of Charles's law and the mathematical expression pertaining to two sets of volume-absolute temperature conditions for a fixed quantity of gas at constant pressure and memorizing it.

Equal Volumes of Gases at Constant Temperature and Pressure Contain Equal Numbers of Molecules: Gay-Lussac and Avogadro

Joseph Gay-Lussac discovered that *volumes of gases always combine in the ratio of small whole numbers to each other, as long as the volumes are measured at the same temperature and pressure.* This statement is now known as Gay-Lussac's **Law of Combining Volumes.** In 1811, Amedeo Avogadro published an explanation for Gay-Lussac's law. Avogadro postulated that *equal volumes of gases under the same conditions of temperature and* *pressure contain equal numbers of molecules.* Thus, 60 mL of H_2(g) combine with 30 mL of O_2(g) to form 30 mL of H_2O(g) at the same temperature and pressure, because the 60 mL of H_2(g) contain twice as many molecules as 30 mL of O_2(g) and H_2(g) and O_2(g) combine in the ratio of two-to-one to form H_2O(g). ✳ Try preparing a "flash card" stating Gay-Lussac's law of combining volumes and memorizing it.

Avogadro's law follows from Avogadro's hypothesis: *The volume of a gas at a given temperature and pressure is proportional to the quantity of gas.* This is why a balloon inflates as air is added. ✳ Try preparing a "flash card" summarizing Avogadro's postulate and law and memorizing it.

12.3 The Ideal Gas Law

The three gas laws considered in 12.2 state that volume is directly proportional to the amount of gas and its temperature and inversely proportional to pressure. The combination of these three laws yields

$$V \propto nT/P \qquad \text{or} \qquad V = R(nT/P) \qquad (12f)$$

where R is a proportionality constant called the **gas law constant**. Rearranging Expression 12f gives the **ideal gas law**

$$PV = nRT \qquad\qquad (12g, 12.4)$$

This equation is called the ideal gas law because it describes "ideal" gases. *Gases at pressures of one atm or less and temperatures around room temperature obey the gas laws rather well and are usually considered to be "ideal" gases.*

The experimental value for the volume occupied by one mole of any gas at the **standard temperature and pressure (STP) conditions of 0°C and 1 atm** is **22.414 L**, the **standard molar volume** of a gas. These values can be used with Expression 12h to calculate **$R = 0.08206$ L·atm/mol·K**. ✳ Try preparing a "flash card" stating the ideal gas law and the units in which P, V, n and T must be expressed for using R = 0,08206 L·atm/mol·K and this value of R and memorizing it. Also try preparing a "flash card" summarizing the standard temperature and pressure conditions for gases and the volume of one mole of any gas, the standard molar volume, at these conditions and memorizing it.

The ideal gas law is useful for calculating one of the four properties of gases when the other three are known. It also can be used to derive the **general** or **combined gas law**

$$\frac{P_1 V_1}{n_1 T_1} = \frac{P_2 V_2}{n_2 T_2} \qquad (12h, 12.5)$$

which can be used to determine the change that will occur when one or more the variables affecting the state of a gas change. Notice how P and V appear on both sides of the combined gas law, so it isn't necessary to convert pressures to atmospheres and volumes to liters to use the combined gas law. **However, it is always necessary to convert temperatures to kelvins to use the gas laws.** ✷ Try preparing a "flash card" stating the combined gas law and memorizing it.

The Density of Gases

Substituting mass (*m*) over molar mass (*M*) for *n* in the ideal gas law and rearranging gives

$$d = \frac{m}{V} = \frac{PM}{RT} \qquad (12i, 12.6)$$

which enables us to calculate the density of a known gas at any given temperature and pressure. In other cases, the density of an unknown gas can be determined at a given temperature and pressure and

$$M = \frac{dRT}{P} \qquad (12j)$$

can be used to calculate the molar mass of the unknown gas. ✷ Try preparing a "flash card" stating the density expression for gases and memorizing it. Also try substituting mass (m) and molar mass (M) into the ideal gas law and deriving the density expression for gases, so you will have something to resort to if you are unable to remember the density expression for gases. Finally, try practicing the algebra for solving Expression 12i for M and obtaining Expression 12j.

Notice that Expression 12i tells us that **the density of a gas is directly proportional to its molar mass.** Air has an average molar mass of 29 g/mol and a density of about 1.2 g/L at 1 atm and 25°C. Thus, **gases having molar masses greater than 29 g/mol have densities greater the 1.2 g/L at 1 atm and 25°C and tend to settle along the ground**. This includes propane, C_3H_8, and butane, C_4H_{10}, and caution must be used when using these common fuels. Indeed, the author is acutely aware of this as a result of an explosion that occurred with a leaking portable cooking stove while on a hike with a Boy Scout troop in the Grand Canyon.

Calculating the Molar Mass of a Gas from P, V and T Data

Solving Expression 12i for molar mass gives

$$M = \frac{mRT}{PV} \qquad \qquad \text{(12k)}$$

and **this expression can be used to determine the molar mass of an unknown gas or of an unknown volatile liquid, as is commonly done in general chemistry laboratory programs. This value can then be used with the simplest formula of the compound to establish the molecular formula of the compound (3.7).** ✳ Try preparing a "flash card" stating the molar mass expression for gases and memorizing it. Also try substituting mass (m) and molar mass (M) into the ideal gas law and deriving the molar mass expression for gases, so you will have something to resort to if you are unable to remember the molar mass expression for gases.

12.4 The Gas Laws and Chemical Reactions

The logic pattern that was developed in Chapter 5 for conducting stoichiometric calculations for chemical reactions can now be extended to include reactions involving gases:

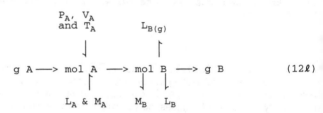

$$\text{(12}\ell\text{)}$$

✳ Try preparing a "flash card" summarizing this logic pattern and memorizing it. Also try using it to solve end-of-chapter stoichiometry problems for chemical reactions.

The ideal gas law tells us

$$n_A = \frac{P_A V_A}{RT_A} \qquad \text{and} \qquad L_B = \frac{n_B RT_B}{P_B} \qquad \text{(12m)}$$

so the mathematical operations for calculating the liters of gas B that can be obtained from a given number of liters of gas A are:

$$\frac{P_A V_A}{RT_A} \cdot \frac{y \text{ mol B}}{x \text{ mol A}} \cdot \frac{RT_B}{P_B} \qquad \text{(12n)}$$

Notice how this expression reduces to

$$V_A \cdot \frac{y \text{ mol B}}{x \text{ mol A}}$$

(12o)

when the gases are at the same pressure and temperature. This is to be expected based on the experimental work of Gay-Lusacc and Gay-Lussac's law. **Keep in mind**, however, **that the logic pattern can be entered at any point and exited at any point, so it can also be used to solve problems involving the reaction of solids to produce gases,** problems like those in Examples 12.9 and 12.10 in the text.

12.5 Gas Mixtures and Partial Pressures

John Dalton was the first to observe that **the pressure of a mixture of gases is the sum of the pressures of the different components of the mixture.** Because the pressure of each individual gas in a mixture is called its **partial pressure**, Dalton's observation is now known as **Dalton's law of partial pressures.**

In a mixture of gases, all of the components necessarily occupy the same volume and have the same temperature. A consequence of this is that **the pressure of one component of a mixture is directly proportional to the number of moles of that component. Thus, for gas A**

$$P_A = \chi_A P_{total} \qquad (12p, 12.7)$$

where χ_A **is the mole fraction of gas A and is given by**

$$\chi_A = \frac{n_A}{n_{total}} = \frac{\text{moles of A}}{\text{total number of moles}} \qquad (12q, 12.8)$$

The sum of the mole fractions are necessarily equal to one for any mixture. ✶ Try preparing a "flash card" summarizing the expressions for calculating the mole fraction of a gas and its partial pressure and memorizing it.

In some general chemistry laboratory programs, nonpolar gases are collected by water displacement. However, some water vapor is usually present in the gas, and it is necessary to solve

$$P_{total} = P_{gas} + P_{water\ vapor} \qquad (12r)$$

for P_{gas} before doing any stoichiometry problems involving the collected gas. The partial pressure of water vapor over liquid water varies with temperature and can be found in a table in Appendix E in the text.

12.6 The Kinetic-Molecular Theory of Gases

Gases are much more compressible than liquids or solids. This implies there is much more space between particles (atoms or molecules) in the gaseous state of matter than in the liquid and solid states of matter. The ability of gaseous molecules to diffuse through each other and spread odors over large distances supports this implication. Furthermore, forces of attraction between gas particles must be nonexistent or small, because gases completely occupy the volume available to them, in contrast to liquids and solids.

It also can be shown that the molecules of a gas move faster as the temperature increases. Indeed, *the average kinetic energy of a collection of gas molecules depends on only the absolute temperature of that collection.* The kinetic energy of an individual molecule is proportional to the square of the speed of the individual molecule (KE = $\frac{1}{2}mu^2$), and *the average kinetic energy for a collection of gas molecules is proportional to the average of the squares of their speeds*, is proportional to their **mean squared speed**. Thus, it can be shown the mean squared speed of the molecules in a collection of molecules is proportional to the absolute temperature of the gas. *That is*,

$$\overline{KE} = \frac{1}{2}m\overline{u^2} = cT \tag{12s}$$

where the line over the KE indicates an average value, the line over the speed squared indicates an average value of the squares of the speeds of the individual molecules, a mean squared speed, and c is a proportionality constant. ✳ Try preparing a "flash card" stating the average kinetic energy expression and memorizing it.

The **kinetic molecular theory** of matter gives a description of the behavior or gases at the atomic or molecular level. *The principal tenets of this theory are:*

• Gases consist of molecules whose separation is much greater than the size of the molecules themselves.

• The molecules of a gas are in continual, random and rapid motion.

• The average kinetic energy of gas molecules is determined by the temperature of the gas. *All molecules regardless of their mass, have the same kinetic energy at the same temperature.*

• Gas molecules collide with one another and with the walls of their container, but they do so without loss of energy.

A gas that strictly obeys all these tenets is said to be an **ideal gas**. In practice, there is no such thing as an ideal gas, but gases do come close to ideal behavior at pressures up to 1 atm and temperatures around room temperature. Assuming ideal behavior, it is possible to derive the ideal gas law using the mathematical statements of the tenets of kinetic-molecular theory. ✶ Try preparing a "flash card" summarizing the principal tenets of the kinetic-molecular theory of gases and memorizing it.

Kinetic-Molecular Theory and the Gas Laws

The gas laws are based on experiment but can be explained by the kinetic-molecular theory. Pressure, for example, is due to collisions with the walls of the container. The higher the temperature of the gas, the higher the number of collisions and more importantly the higher the average force per collision. This is why the P is proportional to T when n and V are constant.

Similarly, increasing the number of molecules of gas in a container increases the number of collisions occurring per second. This is why P is proportional to n when V and T are constant.

On the other hand, if pressure is not allowed to increase when either the temperature is increased or the number of molecules is increased, the area over which collisions can take place must increase through an increase in volume. This is why V is proportional to nT when P is constant in accord with Avogadro's law and Charles's law.

Finally, if the temperature and number of gas molecules is kept constant, the average impact force of the molecules will remain constant but the number of collisions with the walls of the container will increase as the volume of the container is made smaller. Thus, the pressure will increase. This is why P is proportional to $1/V$ when T and n are kept constant in accord with Boyle's law.

Distribution of Molecular Speeds

The number of molecules having a given speed can be determined experimentally, and the number of molecules versus their speed can be plotted, as in the Boltzman distribution curves shown in Fig. 12.13 in the text. The higher the number of molecules having a given speed, the more probable the speed, so the maximum corresponds to the most probable or most common speed.

However, the curves are not symmetrical, because **the average speed is a little faster than the most common speed**.

The higher the temperature, the higher the most common speed, but the curve becomes "flatter" and broader at the higher temperature, as shown in Fig. 12.13 in the text. **Because the average kinetic energy is fixed for all substances at a given temperature, the heavier the gas, the lower its average speed**, as predicted by Expression 12s and shown in Fig. 12.14 in the text. In fact, the square root of the mean squared speed, called the **root-mean-square (rms) speed**, molar mass and absolute temperature are related as

$$\sqrt{\overline{u^2}} = \sqrt{\frac{3RT}{M}} \qquad\qquad (12t, 12.15)$$

where R must be expressed as **R = 8.3145 J/K•mol**. ✶ Try preparing a "flash card" stating the root-mean-square speed expression and memorizing it.

12.7 Diffusion and Effusion

The mixing of molecules of two or more gases due to random molecular motions is called **gaseous diffusion**. Given sufficient time, the molecules of one component of a gaseous mixture will mix thoroughly and completely with all other components of the mixture. Closely related to diffusion is **gaseous effusion** which involves the movement of gas particles through tiny openings in a container.

Thomas Graham discovered experimentally that the rates of effusion of two gases were inversely proportional to the square roots of their molar masses at the same temperature and pressure. Thus,

$$\frac{\text{Rate of effusion of gas 1}}{\text{Rate of effusion of gas 2}} = \sqrt{\frac{\text{molar mass of gas 2}}{\text{molar mass of gas 1}}} \quad (12u, 12.16)$$

This expression, known as **Graham's law, can be used to determine the molar mass of an unknown gas**, as shown in Example 12.15 in the text. ✶ Try preparing a "flash card" stating Graham's law and memorizing it.

12.8 Some Applications of the Gas Laws and Kinetic Molecular Theory

Rubber Balloons and Why They Leak

Rubber balloons have porous surfaces through which gas molecules can effuse. He molecules are lighter than the N_2 and O_2 molecules of air and therefore effuse out of rubber balloons faster than N_2 and O_2 molecules can effuse into the balloons. However, the newer balloons made of Mylar film, which is a polyethylene terephthalate, have smaller pores than rubber balloons and can therefore keep He atoms enclosed much longer.

Deep Sea Diving

The mole fraction of O_2 must be reduced in gases used in deep sea diving, because the partial pressure of oxygen is equal to the mole fraction of oxygen times the total pressure and higher total pressures must be used to combat the pressures experienced at the depths involved. Helium must be used to dilute the breathing gas, because nitrogen is more soluble in blood and body fluids at high pressures and can lead to **nitrogen narcosis**, a condition similar to alcohol intoxication. However, helium atoms are lighter than nitrogen molecules and move with an average speed that is about 2.6 times greater than that of N_2 molecules at the same temperature. Thus, they collide with the body more frequently and are more efficient in carrying heat energy away from the body, so divers often feel chilly when the temperature is a normally comfortable 70°C.

Separation of Uranium Isotopes

The fissionable ^{235}U isotope can be separated from its more abundant ^{238}U isotope by converting the uranium to volatile UF_6 and passing it through thousands of porous membranes. The lighter $^{235}UF_6$ molecules diffuse about 0.4% faster than the heavier $^{238}UF_6$ molecules thereby providing uranium that is enriched from 0.7% ^{235}U in the natural ore to approximately 3% for use as fuel in nuclear power plants and to virtually pure ^{235}U for use in nuclear warheads and atomic bombs.

12.9 Nonideal Behavior: Real Gases

Serious deviations from ideal gas law behavior are observed at pressures higher than one atmosphere and temperatures lower than room temperature. Under these conditions, the volume of the gas is small and the kinetic energy of the gas particles is low, so the assumptions of kinetic molecular breakdown. The volume of the gas particles becomes appreciable in comparison with the total volume of the gas, the volume available to the gas. Furthermore, the particles are closer together and moving more slowly, so forces of attraction between particles, called intermolecular forces, become appreciable and reduce the pressure exerted by the gas particles on the walls of the container.

The so-called **van der Waals's equation**,

$$[P + a(n/V)^2][V - bn] = nRT \qquad (12v, 12.17)$$

developed by Johannes van der Waals, gives more realistic values for pressures than the ideal gas law under such conditions. Because the actual pressure is lower than the ideal pressure calculated by $PV = nRT$ due to intermolecular forces of attraction, the term $a(n/V)^2$ is added to the actual pressure to make this portion of the equation equal to P in $PV = nRT$. The value of the constant "a" can be determined experimentally but has no simple relation to other molecular properties.

The "bn" term is subtracted from the volume of the container to give the free volume available to the gas particles. "b" is an experimentally determined quantity which roughly increases with increasing molecular size. ✳ Try preparing a "flash card" stating van der Waals's equation and memorizing it.

12.10   Key Expressions

Standard atmospheric pressure expression:

   1 atm = 760 mm Hg = 101.325 x 10^3 Pa

Boyle's law: pressure-volume expression:

   $P_1V_1 = P_2V_2$

Charles's law: volume-absolute temperature expression:

   $\dfrac{V_1}{T_1} = \dfrac{V_2}{T_2}$

Ideal gas law expression:

   $PV = nRT$

Combined gas law expression:

$$\frac{P_1 V_1}{n_1 T_1} = \frac{P_2 V_2}{n_2 T_2}$$

Gas density expression:

$$d = \frac{m}{V} = \frac{PM}{RT}$$

Molar mass expressions:

$$M = \frac{dRT}{P} = \frac{mRT}{PV}$$

Logic pattern for stoichiometric calculations:

$$P_A, \; V_A \\ \text{and } T_A \qquad\qquad L_{B(g)}$$

$$\downarrow \qquad\qquad\qquad \uparrow$$

$$g\ A \longrightarrow mol\ A \longrightarrow mol\ B \longrightarrow g\ B$$

$$\uparrow \qquad\qquad \downarrow \qquad \downarrow$$

$$L_A \ \& \ M_A \qquad M_B \quad L_B$$

Partial pressure expression:

$$P_A = \chi_A P_{total}$$

Mole fraction expression:

$$\chi_A = \frac{n_A}{n_{total}} = \frac{\text{moles of A}}{\text{total number of moles}}$$

Average kinetic energy-mean squared speed-absolute temperature expression:

$$\overline{KE} = \tfrac{1}{2}m\overline{u^2} = cT$$

Root-mean-square (rms) speed expression:

$$\sqrt{\overline{u^2}} = \sqrt{\frac{3RT}{M}}$$

Graham's law expression:

$$\frac{\text{Rate of effusion of gas 1}}{\text{Rate of effusion of gas 2}} = \sqrt{\frac{\text{molar mass of gas 2}}{\text{molar mass of gas 1}}}$$

van der Waals's equation expression:

$$[P + a(n/V)^2][V - bn] = nRT$$

12.10   Hints for Answering Questions and Solving Problems

Working with the gas laws:

Remember to:

- use absolute temperatures with all gas laws involving temperature.

- use tables summarizing initial and final condition values before substituting into Boyle's law, Charles's law and the Combined gas law. (See Examples 12.2, 3 and 6 in the text.)

- convert pressures to atmospheres and volumes to liters before substituting into the Ideal gas law when using $R = 0.08026$ L·atm/mol·K. (See Ex. 12.5 in the text.)

Giving explanations:

Try writing short paragraphs to explain how the kinetic-molecular theory of gases accounts for the properties of gases and the gas laws. Also try writing a short paragraph to explain why gases deviate from "ideal" behavior at high P and low T.

CHAPTER 13

BONDING and MOLECULAR STRUCTURE:
INTERMOLECULAR FORCES, LIQUIDS AND SOLIDS

The behavior of gases was discussed in Chapter 12. This chapter focusses on the characteristics and behavior of liquids and solids and the nature of the intermolecular forces of attraction that cause substances to exist as liquids and solids. Almost all gases behave similarly whereas *the behavior of liquids and solids depends on the composition and structures of the substances involved, because composition and structure affect intermolecular forces of attraction.*

13.1 States of Matter and the Kinetic-Molecular Theory

The molecules of liquids and solids are much closer together than those of gases, and there are significant intermolecular forces of attraction holding the molecules together. Thus, liquids and solids are not as compressible as gases, and liquids and solids do not expand to completely and uniformly fill their containers like gases. The molecules are still in motion and their kinetic energy is still determined only by temperature, but their motion is restricted by the intermolecular forces.

13.2 Intermolecular Forces

The attractive forces between molecules or between ions and molecules are called **intermolecular forces.** There are four main types of intermolecular forces: *ion-dipole, dipole-dipole (including hydrogen bonding), dipole-induced dipole and induced dipole-induced dipole.* The characteristics of these forces are summarized in Table 13.1 in the text. *Notice that intermolecular forces are weaker than the forces between ions that hold ionic compounds together and weaker than the covalent bonds between atoms that hold atoms together within molecules.* ✳ Try preparing a "flash card" listing the four main types of intermolecular forces, the principal factor responsible for the interaction energy for each and the approximate range of energy associated with each force and memorizing it.

Interactions Between Ions and Molecules with a Permanent Dipole

A dipole consists of separated positive and negative charges (9.6), so an ion can be attracted to one end of the dipole and be repelled by the other. The strength of an ion-dipole attraction depends on:

- *the distance between the ion and the dipole*: the closer the ion and the dipole, the stronger the attraction.

- *the charge on the ion*: the higher the ion charge, the stronger the attraction.

- *the magnitude of the dipole*: the greater the magnitude of the dipole, the stronger the attraction.

The force of attraction between an ion and a dipole actually depends on $1/d^2$, where d is the distance between the center of the ion and the end of the dipole to which it is attracted. *The ion-dipole force is weaker than the force of attraction between ions of opposite charge but stronger than any other intermolecular force.* ✳ Try preparing a "flash card" summarizing the factors affecting the strength of ion-dipole interactions and memorizing it.

An ion-dipole interaction that is of particular interest to chemists involves the interaction of ions with water molecules, because these interactions are involved in the dissolving of ionic solids in water. An ion bound to water molecules is said to be **hydrated**, and the energy change for the process is called the **heat**, or **enthalpy, of hydration.** *In comparing ions of equal charge, the larger the ion radius, the less exothermic the enthalpy of hydration.* In cases where the interaction with water is sufficiently strong, water molecules actually remain with the metal cation when the salt is collected from solution by evaporation, and a **hydrated salt** is obtained. In the case of ions bound to other polar solvents, the ion is said to be **solvated**, and we use the term **enthalpy of solvation**.

Interactions Between Molecules with Permanent Dipoles

Molecules containing atoms of different electronegativity bonded together unsymmetrically contain permanent dipoles (9.6). The positive end of the dipole of one molecule is attracted to the negative end of the dipole of another molecule, as shown in Fig. 13.5 in the text. *These attractions affect melting and boiling points, because molecules are separated from one another when a solid melts or a liquid boils. In general, the melting point and boiling point for a polar compound are higher than those for a nonpolar compound of approximately equal molar mass.*

The presence of a permanent dipole also influences solubility. Polar molecules dissolve in polar solvents, and nonpolar molecules dissolve in nonpolar solvents giving rise to the saying "like dissolves like."

Hydrogen Bonding

When a hydrogen atom is bonded to a small, highly electronegative atom X interacts with a second small, highly electronegative atom X or Y, with X and Y restricted to N, O or F, the dipole-dipole interaction is particularly strong and is called **hydrogen bonding.**

$$\overset{\delta_-}{X} - \overset{\delta_+}{H} \cdots \overset{\delta_-}{Y} \qquad (13a)$$

Hydrogen bonding can occur between molecules (intermolecular hydrogen bonds) or, if the geometry allows, as in DNA, within molecules (intramolecular hydrogen bonds). *The effect of hydrogen bonding is apparent in the high boiling points of NH_3, H_2O and HF,* as shown in Fig. 13.8 in the text.

The Unusual Properties of Water: Consequence of Hydrogen Bonding

The boiling point of water is almost 200°C higher than expected based on the extrapolation of the line established by H_2S, H_2Se and H_2Te. This is due to the optimum combination of two H atoms and two lone pairs of electrons about the O atom of H_2O leading to a maximum of four hydrogen bonds to four other water molecules. As water freezes, these hydrogen bonds become more rigid yielding the more open structure that is shown in Fig. 13.10 in the text, and this causes ice to be less dense than liquid water, whereas the solid state of most substances is more dense than the corresponding liquid state.

Other anomalous properties of water are summarized in Table 13.4 in the text. *The extraordinarily high heat capacity of water is related to its hydrogen bonding and is of particular interest to us because it influences the ability of bodies of water to moderate temperatures.*

Dispersion Forces: Interactions Involving Induced Dipoles

Dispersion forces are found in all substances. Such forces are electrostatic in nature and arise from attractions involving induced dipoles. *They can be used to explain why atoms of the noble gases and nonpolar molecules of compounds can be converted to liquids and/or solids.*

Interactions Between Polar and Nonpolar Molecules:

O_2 is nonpolar and isn't expected to be very soluble in water, which is highly polar. The very fact that it dissolves in water at all ($\approx$ 0.001% by weight) is due to the ability of polar molecules to induce a dipole in a nonpolar molecule, as shown in Fig. 13.14 in the text. *The approach of the polar molecule causes a shift in the electron density of the nonpolar molecule thus inducing a dipole in the otherwise nonpolar substance.*

The process of inducing a dipole is called **polarization**, and the degree to which the electron cloud of an atom or molecule can be distorted depends on its **polarizability**. *In general, the more electrons there are in an atom or molecule, the more easily the electrons can be polarized, because they are further from the restraining forces of atomic nuclei.*

Polarizability therefore increases with increasing size, and increasing size is usually accompanied by increasing molar mass. Thus, the solubility of nonpolar gases in sea water increases with both increasing size and increasing molar mass as shown in Table 13.5 in the text. *Keep in mind, however, that use of molar mass to predict the relative strengths of dispersion forces is a matter of correlation, rather than explanation. Mass only affects gravitational attractions between large objects, and gravitational forces between objects as small as atoms and molecules are virtually nonexistent. Nevertheless, it is convenient to correlate the relative strengths of dispersion forces with molar mass and to use the polarizability of atoms and molecules to give explanations for these relative strengths.*

Interactions Between Nonpolar Molecules:

Nonpolar molecules are attracted to one another through induced dipole-induced dipole forces, which arise from momentary distortions in electron clouds and can range from weak to relatively strong. There is a clear trend toward higher boiling points of nonpolar gases with increasing molar mass, because dispersion forces generally become stronger with increasing size and polarizability. This means a higher temperature is required to overcome the forces of attraction between the molecules to enable them to enter the vapor phase. It is also means that more energy is needed to separate the molecules from one another at the boiling point, and this is reflected in more endothermic enthalpies of vaporization with increasing size; see Table 13.6 in the text. Also see the illustration of the interaction between induced dipoles in Fig. 13.15 in the text.

Determining What Types of Intermolecular Forces are Operative in a Given Chemical System

Fig. 13.6 in the text gives a scheme that can be used to decide what type of intermolecular forces are operating in a given chemical system. *Keep in mind, however, that dispersion forces are found in all systems*.

13.3 Properties of Liquids

Gases condense to liquids when the molecules no longer have sufficient kinetic energy to overcome the intermolecular forces. In this section, the properties of liquids are considered.

Enthalpy of Vaporization

The molecules in the liquid state of a substance have a distribution of energies (Fig. 13.7 in the text) *that resembles the distribution of speeds/energies for gas molecules* (Fig. 12.13 in the text). *As with gases, the average kinetic energy of the molecules depends only on temperature: the higher the temperature, the higher the average kinetic energy and the greater the relative number of molecules with high kinetic energy.*

If the molecules of a liquid

- have sufficient kinetic energy to overcome the potential energy of the intermolecular forces of attraction holding the liquid together and

- are at the surface of the liquid and moving in the right direction,

they can enter the gaseous state in the process called **vaporization**, or **evaporation**. However, *as they leave and take their kinetic energy with them, there is a lowering of the average kinetic energy, and the temperature of the molecules remaining in the liquid state is lowered, as well. Thus, an input of energy is required to produce more molecules with kinetic energy greater than the minimum for escape and continue the process*. Perhaps you've experienced this as swimmer's chill on emerging from water or are aware of the use of body rubs with isopropyl alcohol to reduce the temperature of people who are ill and have a fever.

The heat energy that is required to overcome the attractive forces and vaporize a liquid is called its **heat of vaporization,** expressed in joules per gram, or its **molar enthalpy of vaporiza-**

tion, $\Delta H_{\text{vaporization}}$, expressed in kJ per mole. In closed containers, gas molecules reenter the liquid phase in the exothermic process called **condensation**. *The enthalpy change for condensation is equal but opposite in sign to the enthalpy change for vaporization:*

$$\Delta H_{\text{condensation}} = -\Delta H_{\text{vaporization}} \tag{13b}$$

The amount of heat energy involved in either process can be calculated by multiplying the number of moles being changed by the $\Delta H_{\text{vaporization}}$ *or the* $\Delta H_{\text{condensation}}$, as in Ex. 13.3 in the text.

The ΔH_{vap} *and boiling point values for nonpolar liquids, such as the hydrocarbons, atmospheric gases and the halogens, increase with increasing size and molar mass as a reflection of increasing intermolecular dispersion forces*, as shown in Table 13.6 in the text. *Similar trends are observed for the heavier hydrogen halides, HCl, HBr and HI. Here the increase in dispersion forces with increasing size and molar mass offsets the decrease in dipole-dipole forces from HCl to HI whereas the corresponding values for HF are higher than expected due to hydrogen bonding.*

Vapor Pressure

Liquids placed in closed containers attain a state of **dynamic equilibrium** with their vapors. *Molecules continue to leave the liquid and enter the vapor phase as molecules reenter the liquid phase at the same rate.*

$$\text{Liquid} \rightleftarrows \text{Vapor} \tag{13c}$$

The pressure exerted by the vapor at equilibrium is called the **equilibrium vapor pressure** of the substance. *The vapor pressure of any substance is a measure of the tendency of its molecules to leave the liquid phase and enter the vapor phase at a given temperature.* We refer to this tendency as the **volatility** of the compound. The more **volatile** the compound, the higher the equilibrium vapor pressure at a given temperature.

As temperature increases, the molecules of the liquid move more rapidly causing the rate of vaporization to increase. The equilibrium vapor also increases, so that any given point along a vapor pressure versus temperature curve for a compound the liquid and vapor are in equilibrium. R. Clausius and B.P.E. Clapeyron first showed that the *vapor pressure is related to temperature by the Clausius-Clapeyron equation*

$$\ln P = -\Delta H_{\text{vap}}/RT + C \tag{13d}$$

The term ln P represents the natural logarithm of the vapor pressure, ΔH_{vap} is the enthalpy of vaporization for the liquid,

T is the Kelvin temperature at which P is measured, R is the ideal gas constant expressed as 8.314 J/mol·K and C is a constant characteristic of the compound involved. This equation is of the slope-intercept form and thus tells us that the value of ΔH_{vap} can be obtained by measuring P at various temperatures and plotting ln P vs. $1/T$. Furthermore, once the value of ΔH_{vap} is known, the value of C can be determined by using the value of P at any given temperature, and the value of P can then be calculated at any other temperature.

Boiling Point

Bubbles of vapor begin to form in a liquid as it is heated. The **boiling point** of a liquid is the temperature at which the vapor pressure of the liquid is equal to the external pressure, and if the external pressure is 1 atm, the temperature is designated as the **normal boiling point** of the liquid. The normal boiling point of water is 100°C. However, *the boiling point decreases with increasing elevation and decreasing atmospheric pressure*, so foods take longer to cook at elevations higher than sea level. On the other hand, foods take less time to cook in pressure cookers, because the prevailing pressure is greater than atmospheric pressure and water boils at a higher temperature.

There is a direct relation between normal boiling point, molar enthalpy of vaporization and intermolecular forces. The stronger the intermolecular forces, the higher the normal boiling point and the more endothermic the molar enthalpy of vaporization.

Critical Temperature and Pressure

The vapor pressure of a liquid continues to increase with temperature up to a point called the **critical point**, where it comes to an abrupt halt. The temperature at which this occurs is the **critical temperature, T_c,** and the corresponding pressure is the **critical pressure, P_c.** At temperatures greater than T_c, the molecules of the liquid possess sufficient kinetic energy to enter the vapor phase, regardless of the pressure, and a conventional liquid ceases to exist. Instead a supercritical gas is formed. A **supercritical gas** is one at a pressure so high that its density resembles that of a liquid, while its **viscosity** (ability to flow) remains close to that of a gas.

Air conditioners and refrigerators work on the principle of removing heat by vaporizing a liquid and compressing the vapor at $T < T_c$ and $P > P_c$ to condense the vapor to the liquid state, so it can be recycled. Propane and butane are used in portable grills and home heating and cooking because they have relatively high critical temperatures and can easily be liquefied. In addition, supercritical nonpolar CO_2 is being used to remove nonpolar oils from foods and caffeine from coffee. It has an advantage over nonpolar liquids, because it is easily vaporized and does not leave residual traces of the extracting material as liquids do.

Surface Tension, Capillary Action and Viscosity

Molecules on the surface of a liquid are only attracted inward by those below the surface layer. Thus, the surface contracts and acts as though it has a "skin," the toughness of which is measured by its **surface tension,** by the energy required to break through the surface. *Surface tension is the reason certain objects that are more dense than water can be made to float on the surface of water and the reason certain insects are able to walk on water. It is also the reason why water droplets form spheres, because a sphere has a smaller surface than any other shape of the same volume.*

Capillary action occurs when a liquid is placed in a tube of small diameter. When water is placed in a glass tube, **adhesive forces** of attraction between the polar water molecules and the polar Si-O bonds of the surface of the glass exceed the **cohesive forces** of attraction between molecules. This causes the characteristic concave, or downward-curving, **meniscus** that is seen with water in test tubes or water solutions in burets. By contrast, mercury exhibits a convex, or upward curving, meniscus, because its cohesive forces (high surface tension forces) are stronger than the adhesive forces between mercury and glass.

Viscosity refers to resistance to flow and is influenced by both molecular shape and intermolecular forces. Long chain molecules, like those of natural oils, become entangled with one another causing higher viscosities. Also, longer chains experience greater intermolecular forces, because there are more atoms to attract one another. *Viscosity generally decreases with increasing temperature because the increase in kinetic energy overcomes the intermolecular forces.* This is why automobiles require the use of "multi-weight", multi-viscosity, motor oils.

13.4 Metallic and Ionic Solids

Solids can be classified as ionic, metallic, molecular, network or amorphous. The characteristics of each of these types is given in Table 13.8 in the text. ✶ Try preparing a "flash card" summarizing this information and memorizing it.

Crystal Lattices and Unit Cells of Metal Atoms

Crystalline metals and compounds contain atoms, ions, or

molecules arranged in orderly **crystal lattices**. All crystal lattices are built of **unit cells**, which are the smallest, repeating unit that has all of the symmetry characteristics of the overall structure. ***Thus, the external appearance of a crystalline solid is a reflection of its unit cell.***

The unit cells that are of most interest to us are **cubic unit cells**, cells with edges of equal length that meet at 90° angles. There are three different types of cubic unit cells: **simple cubic (sc)**, **body-centered cubic (bcc)** and **face-centered cubic (fcc)**. *All three have eight identical atoms or ions at the corners of cube. A bcc unit cell has one additional atom at the center of the cube whereas a fcc unit cell has additional atoms in each of the six faces of the cube.* These arrangements are illustrated in Fig. 13.32 in the text. ✳ Try preparing a "flash card" for each type of cubic unit cell showing the arrangement of atoms for each type and memorizing the pictures and the word descriptions for each type.

An atom or ion at the corner of a cubic unit cell is actually part of eight unit cells (Fig. 13.33a in the text), so only $\frac{1}{8}$ of a corner atom or ion is within a given unit cell. The result is that there is one net atom or ion ($8 \times \frac{1}{8}$) within the unit cell, so a simple cubic arrangement has only one net atom or ion per unit cell. A bodycentered arrangement has the additional atom at the cube center that is wholly within the unit cell, so a body-centered unit cell has a net of two atoms or ions per unit cell. Lastly, each atom in the face of a face-centered unit cell is part of two unit cells, so there is a net of three ($6 \times \frac{1}{2}$) atoms per unit cell due the these positions, and a face-centered unit cell has a net of four atoms or ions per unit cell. Ex. 13.5 in the text shows how this information can be combined with other experimental information to calculate the radius of an atom.

Structures and Formulas of Ionic Solids

The structures of many ionic compounds can be built by taking a simple cubic or face-centered cubic lattice of spherical ions of one type and placing ions of opposite charge in holes left in the lattice. The hole in a simple cube is at the center of the unit cell, and this is the only type of hole available. CsCl exists in a structure which has Cl^- ions at the corners of the cubes and a Cs^+ ion at the center of each cube.

The structure of NaCl consists of a face-centered arrangement of Cl^- ions with Na^+ ions in each of the octahedral holes. These holes are located at the cube center and along each of the 12 edges of the cube. The holes are called octahedral holes because each Na^+ is surrounded by 6 Cl^- ions at the corners of an octahedron, as shown in Fig. 13.34 in the text.

The ratio of ions in the unit cell must be compatible with the formula of the compound. In the case of NaCl, there are 4 Na^+ and 4 Cl^- ions per unit cell for a 1 Na^+:1 Cl^- ratio. *These numbers can be calculated by using the number of ions of each type in the unit cell and the fraction of ions in each position belonging to the unit cell. For NaCl this gives:*

$$(1 \text{ cube center} \times 1) + (12 \text{ cube edges} \times \tfrac{1}{4}) = 4 \ Na^+ \text{ ions} \qquad (13e)$$

$$(8 \text{ cube corners} \times \tfrac{1}{8}) + (6 \text{ cube faces} \times \tfrac{1}{2}) = 4 \ Cl^- \text{ ions} \qquad (13f)$$

There is always a net of 4 octahedral holes in a fcc arrangement and a net of 4 positions occupied by the ions forming the fcc arrangement.

Summary:

Ionic compounds having the formula MX commonly exist in the CsCl or NaCl (rock salt) structures. In the

- *CsCl structure*, the M^{n+} cations occupy cubic holes in a simple cubic X^{n-} anion lattice and are each surrounded by 8 anions.

- *NaCl structure*, the M^{n+} cations occupy all of the octahedral holes in a face-centered cubic X^{n-} lattice and are each surrounded by 6 anions.

The alkali metal halides, except CsCl, CsBr and CsI, all the oxides and sulfides of the alkaline earth metals and all the oxides of transition metals of the fourth period having the formula MO occur in the rock salt structure. *Only the larger Cs^+ commonly forms MX compounds having the CsCl structure, because other cations are too small to be surrounded by eight anions without having the anions trying to occupy the same space.*

The formula of an ionic compound can always be derived from the unit cell structure. Furthermore, because metallic or ionic solids form by packing atoms or ions as closely as possible, *the size of the unit cell and the density of the solid depend on the radii of the atoms or ions involved.* Exs. 13.5, 6, 7 and 8 in the text illustrate important types of calculations based on unit cells, and Problem Solving Tips 13.1 and 13.2 give important information that can be used to solve such problems.

13.5 Molecular and Network Solids

Molecular Solids

Covalently bonded molecules, such as I_2, CO_2 and H_2O, condense to form solids as the temperature and kinetic energy of molecules is lowered to the point where the intermolecular forces are strong enough to keep the molecules in place. In the case of nonpolar molecules, like I_2 and CO_2, *the only intermolecular forces operating are dispersion forces. In the case of polar molecules, dispersion forces and dipole-dipole forces are operating, and in cases where H is covalently bonded to N, O or F and is attracted to another N, O or F atom (Expression 13a), the dipole-dipole forces are strong enough to be called hydrogen bonds.* The structure of ice, which is an important molecular solid, was described in Section 13.2 (Fig. 13.10) in the text.

Network Solids

A number of solids are composed of networks of covalently bonded atoms. Graphite and diamond are two network solids that are discussed in Section 3.1 in the text. Diamonds can be made by heating graphite to temperatures of 1500°C in the presence of a metal, such as iron or nickel, at pressures of 50,000 to 65,000 atm. Diamonds can also be made by heating a mixture of hydrogen and a carbon-containing compound like methane (CH_4) to 2200°C. Carbon atoms deposit on the surface of a silicon plate or other material in a process called "chemical vapor deposition" (CVD) and slowly build up a film of tiny diamonds. The properties of diamonds are listed on p. 637 in the text. ✱ Try preparing a "flash card" listing the properties of diamonds and memorizing it.

Silicates are composed of silicon and oxygen and are another important class of network solids. You know them as sand, quartz, talc and mica. Silicate structures are described in Section 22.5 in the text.

13.6 The Physical Properties of Solids

The **melting point** of a solid is the temperature at which the solid is converted to liquid. The heat energy that is required to overcome the attractive forces to melt a solid is called its **heat of fusion**, expressed in joules per gram, or its **enthalpy of fusion, ΔH_{fusion},** expressed in kJ per mole. *The enthalpy change for freezing is equal but opposite in sign to the enthalpy* *change for melting:*

$$\Delta H_{\text{crystallization}} = -\Delta H_{\text{fusion}} \qquad (13g)$$

A substance that has weak intermolecular forces generally has a low enthalpy of fusion and a low melting point whereas a substance with strong intermolecular forces generally has a high enthalpy of fusion and a high melting point. In general then, low-molecular weight, nonpolar substances that form molecular solids have low melting points. However, the melting points increase as the size and molar mass increase, because the dispersion forces become stronger with increasing size.

Ionic compounds always have higher melting points than molecular solids due to strong cation-anion forces. Indeed, because lattice energies depend on ion charges and ion sizes, as reflected in the distance between ions (8.7), *there is a good correlation between the lattice energy of alkali metal halides and the position of the metal or halogen in the periodic table:*

• As the alkali metal cation size increases from Li^+ to Cs^+, there is a decline in the lattice energy of compounds of a given halide anion.

• As the halide anion size increases from F^- to I^-, there is a decline in the lattice energy of compounds of a given alkali metal cation.

These trends are illustrated in Fig. 13.39 in the text, and *the data given for the sodium halides in Table 13.9 in the text show the decrease in melting point and enthalpy of fusion values that occurs with increasing halide anion size and decreasing lattice energy.* ✶ Try preparing a "flash card" summarizing the information of Fig. 13.39 in the text and its relation to the melting points and enthalpies of fusion of the alkali halides and memorizing it.

In some cases, molecules can move directly from the solid to the gas phase by **sublimation**. The heat energy that is required to overcome the attractive forces to sublime a solid is called its **enthalpy of sublimation, $\Delta H_{\text{sublimation}}$.** The reverse process is called **deposition**, and *the enthalpy change for deposition is equal but opposite in sign to the enthalpy change for sublimation:*

$$\Delta H_{\text{deposition}} = -\Delta H_{\text{sublimation}} \qquad (13h)$$

Water has a low enthalpy of sublimation (51 kJ/mol), and the freeze-drying of foods process and "frost-free" refrigerators take advantage of this. Iodine and carbon dioxide are other common substances that sublime.

Ex. 13.9 in the text illustrates calculating the amount of heat energy that is required to change the temperature and physical state of a substance. **Recall from Chapter 6 that the heat energy that is associated with a temperature change is given by**

$$q = \text{(specific heat)(mass)(temperature change)} \qquad (13i)$$

and that associated with a change of state is given by

$$q = \text{(moles)(change of state enthalpy change in kJ/mol)} \qquad (13j)$$

13.7   Changes in Structure and Phase

Phase Diagrams

Phase diagrams are used to illustrate the relation between the phases of matter and pressure and temperature conditions. **Each point along a line in a phase diagram represents a P and T at which an equilibrium exists between the two phases on either side of the line.** The intersection of the solid-liquid, solid-vapor and liquid-vapor equilibrium lines occurs at the **triple point**, the point at which all three phases (solid, liquid and vapor) can exist in equilibrium.

The triple point of water occurs at $P = 4.58$ mmHg and $T = 0.01°C$, as shown in Fig. 13.41 in the text. **Notice how the dashed line across the phase diagram of water at 760 mmHg intersects the solid-liquid equilibrium line at 0°C and the liquid-vapor equilibrium line at 100°C, the normal melting and boiling points of water. Also notice how the solid-liquid equilibrium line slants to the left indicating the melting point of water decreases with increasing pressure.** It actually changes by only 0.01°C for every 1 atm increase in pressure.

Water is one of only three substances known to have solid-liquid equilibrium lines with negative slopes; the others are antimony and bismuth. **The negative slope for the solid-liquid equilibrium line for water can be rationalized by realizing that ice has a more open and less dense structure than liquid water as a result of more directional and more rigid hydrogen bonding. Thus, an increase in pressure should favor a decrease in volume and produce the more dense liquid phase of water.** Indeed, the pressure caused by the blade of an ice skate is sufficient to lower the melting point of ice by about 5°C and combine with frictional heating to enable a skater to skate on a thin film of liquid water.

The phase diagram for carbon dioxide is shown in Fig. 13.43 in the text. *Notice that the entire liquid phase region for carbon dioxide is above 5.2 atm, so it is impossible for carbon dioxide to melt at 1 atm. Instead, it sublimes at temperatures above $-78°C$ at 1 atm.* This is why solid carbon dioxide is known as "dry ice"; it actually cools substances as it removes heat to become a vapor and does so without becoming a liquid.

13.8 Key Expressions

Relations among molar enthalpy changes for changes of state:

$$\Delta H_{condensation} = -\Delta H_{vaporization}$$

$$\Delta H_{crystallization} = -\Delta H_{fusion}$$

$$\Delta H_{deposition} = -\Delta H_{sublimation}$$

Vapor pressure-temperature relation expression:

$$\ln P = -\Delta H_{vap}/RT + C$$

Heat flow for temperature change expression:

$$q = (specific\ heat)(mass)(temperature\ change)$$

$$= (c\ in\ J/g \cdot K)(mass\ in\ grams)(\Delta T\ in\ kelvins)$$

Heat flow for change of state expression:

$$q = (mass\ in\ grams)(change\ of\ state\ heat\ flow\ in\ J/g)$$

$$= (moles)(change\ of\ state\ enthalpy\ change\ in\ kJ/mol)$$

13.9 Hints for Answering Questions and Solving Problems

Correlating the content of the chapter:

The interrelations between the topics discussed in this chapter are shown in Fig. 13.1 in the text. ✱ Try preparing a "flash card" showing these relations and memorizing it.

Identifying types of intermolecular forces for chemical systems:

A flow diagram for identifying types of intermolecular forces is given in Fig. 13.16 in the text. ✶ Try preparing a "flash card" showing this information and memorizing it, but also keep in mind that dispersion forces occur with all chemical systems.

Correlating intermolecular forces with physical properties:

Substances having low intermolecular forces of attraction generally have:

- low melting points and low enthalpies of fusion,
- low surface tension and are not viscous, and
- high vapor pressures, low boiling points and low enthalpies of vaporization.

Calculating the number of atoms or ions in a given unit cell:

The key to calculating the number of atoms or ions in a given unit cell is to know the type of locations of the atoms or ions, the number of locations of each type in the unit cell and the fraction of each type of site that belongs to the unit cell. This information can be summarized as:

Location description	Positions of atoms or ions	Number of positions	Fraction of atom or ion in unit cell
simple cube, sc	cube corners	8	$\frac{1}{8}$
body-centered cube, bcc	cube corners	8	$\frac{1}{8}$
	cube center	1	1
face-centered cube, fcc	cube corners	8	$\frac{1}{8}$
	cube faces	6	$\frac{1}{2}$
cubic hole in sc	cube center	1	1
octahedral holes in fcc	cube center	1	1
	cube edges	12	$\frac{1}{4}$

Applying this information to unit cells gives number of:

- atoms in simple cube = 8 corners$\cdot\frac{1}{8}$ = 1.

- atoms in bcc = (8 corners$\cdot\frac{1}{8}$) + (1 center$\cdot$1) = 2.

- atoms or ions in fcc = (8 corners$\cdot\frac{1}{8}$) + (6 faces$\cdot\frac{1}{2}$) = 4.

- ions in cubic hole in sc = 1 center·1 = 1.

- ions in octahedral holes in fcc = (1 center·1) + (12 edges·¼) = 4.

✳ Try preparing a "flash card" for the CsCl structure stating the description of the unit cell (p. 172), the location and number of ions of each type and the fraction of each ion belonging to the unit cell and summarizing the procedure for calculating the number of ions of each type in the unit cell and memorizing it. Also try preparing a "flash card" giving the same information for the NaCl structure and memorizing it.

Conducting calculations based on unit cells:

A logic pattern for conducting calculations based on unit cells is given in "Problem Solving Tips and Ideas" 13.2 in the text, and comments are also given on how this pattern is used in solving Exs. 13.5, 6 and 8 in the text. ✳ Try preparing a "flash card" summarizing this logic pattern and a separate "flash card" for each of these examples outlining the information given, the goal of the example and the steps to be followed to solve the problem, as outlined in the strategy portions of the solutions for these examples, and reviewing this information until you are comfortable conducting these types of calculations.

CHAPTER 14

SOLUTIONS and THEIR BEHAVIOR

A **solution** is a homogeneous mixture composed of two or more substances in a single phase. The component that is present in the largest amount is called the **solvent** and the other component is the **solute**. The solvent can be a solid, liquid or gas, but this chapter focusses on the concentrations and properties of solutions having liquids as solvents. Preparing a solution often changes the properties of the solvent. Properties, such as vapor pressure, boiling point, freezing point and osmotic pressure, ideally depend on only the number of solute particles per particle of solvent and not on the nature of the solute or solvent and are called **colligative properties.** Solutions are an important part of life, so it is important to understand the nature and properties of solutions.

14.1 Units of Concentration

Molarity cannot be used for colligative property calculations, because there is no way of knowing the number of solvent molecules that are used to prepare a solution of given volume by diluting to a fixed mark, as shown in Fig. 14.1 in the text. *Thus, units such as molality, mole fraction, weight percent and parts per million must be used with colligative properties.*

• The **molality** of a solution is defined as the number of moles of solute per kilogram of solvent.

$$\text{Molality of A } (m) = \frac{\text{moles of A}}{\text{kilograms of solvent}} \qquad (14a, 14.1)$$

The molality of an aqueous solution is approximately equal to the molarity of the solution when the molarity is less than 0.01 M, because, under these conditions, *the volume of solute plus volume of one kilogram of water is essentially equal to the volume of one liter of solution*. In the case of the 0.100 m and 0.100 M K_2CrO_4 aqueous solutions shown in Fig. 14.1 in the text, the volume is clearly not the same, so 0.100 m K_2CrO_4 is not the same as 0.100 M K_2CrO_4.

• The **mole fraction** of component A, χ_A, is defined as the number of moles of A divided by the total number of moles

---

of all of the components present (12q, 12.8). For a solution composed of one solute A and water,

$$\chi_A = \frac{\text{moles of A}}{\text{moles of A + moles of water}} \qquad (14b, 14.2)$$

Because **the sum of the mole fractions are necessarily equal to one for any mixture**,

$$\chi_A + \chi_{water} = 1.00 \qquad (14c)$$

• The **weight percent** of component A is equal to the mass of A divided by the total mass of solution times 100%.

$$\text{Wt.\% A} = \frac{\text{grams of A}}{\text{grams of solution}} \times 100\% \qquad (14d)$$

For a solution composed of one solute A and water, this becomes

$$\text{Wt.\% A} = \frac{\text{grams of A}}{\text{grams of A + grams of water}} \times 100\% \qquad \begin{matrix}(14e, \\ 14.3)\end{matrix}$$

• The **parts per million** of component A, ppm A, is defined as the milligrams of component A per liter of solution for aqueous solutions.

$$\text{ppm A} = \frac{\text{milligrams of A}}{\text{L of solution}} \qquad (14f)$$

Exs. 14.1 and 14.2 in the text illustrate calculations involving these units. ✱ Try preparing a "flash card" for each of these concentration unit definitions and memorizing them. Also try practicing problems involving these units until you feel comfortable using them and can do them quickly.

14.2 The Solution Process

When a solute dissolves in a solvent, the solute-solvent attractive forces must be great enough to overcome the intermolecular forces between the pure solvent molecules and the intermolecular forces within the solute. The solute-solvent forces can cause the solute to become solvated if the solvent molecules are sufficiently attracted to the solute molecules or ions. When water is the solvent, the solvation is called hydration (13.2).

Solubility refers to the maximum amount of material that can dissolve in a given amount of solvent at a given temperature to produce a stable solution. Qualitative guidelines for the solubility of ionic compounds were given in Section 4.3, but the objective here is to be more explicit. **When**

- as much solute is dissolved as can be dissolved under the conditions, the solution is **saturated.**

- less solute is dissolved than can be dissolved under the conditions, the solution is **unsaturated**.

- more solute is dissolved than can be dissolved under the conditions, the solution is **supersaturated.** (Supersaturated solutions are prepared under other conditions and are unstable under their present conditions; see the caption for Fig. 14.2 in the text.)

To put these terms in perspective, note that the text states that 950 g of $AgNO_3$ will dissolve in 100 mL of water at 100°C but only 0.00217 g of AgCl will dissolve in 100 mL of water at 100°C. *We say that $AgNO_3$ is soluble, and AgCl is insoluble, though AgCl dissolves to a small degree. We also say that both solutions are saturated, but the solution of AgCl is a dilute solution* ($\approx 1.5 \times 10^{-4}$ M) *and the solution of $AgNO_3$ is a concentrated solution* (≈ 56 M).

Liquids Dissolving in Liquids

If two liquids mix to an appreciable extent to form a solution, they are said to be **miscible**. In contrast, two liquids that do not mix to form a solution but exist as layers in contact with one another are said to be **immiscible**. Polar ethanol (C_2H_5OH) molecules dissolve in all proportions in water whereas nonpolar carbon tetrachloride (CCl_4) molecules and water are immiscible. On the other hand, nonpolar carbon tetrachloride molecules dissolve in all proportions in nonpolar octane (C_8H_{18}). These observations and numerous others have led to the guideline: *like dissolves like*. ✳ Try preparing a "flash card" stating this guideline and summarizing the meaning of it and memorizing it.

The enthalpy change for forming a solution of two liquids, $\Delta H_{solution}$, can be analyzed by breaking the process down into a series of steps, as shown on p. 660 in the text. In this cycle,

- the solute and solvent liquids are both vaporized in endothermic steps,

• the vaporized solute and solvent molecules form intermolecular associations in an exothermic step as a result of intermolecular forces of attraction, and

• the solute-solvent mixture condenses in an exothermic step.

The overall solution process is usually exothermic when the solute-solute, solvent-solvent and solute-solvent forces are the same type, and the process is product-favored. ✳ Try preparing a "flash card" summarizing the energy cycle for dissolving a liquid in a liquid and memorizing it. Use solute liquid for one of the reactants and solvent liquid for the other to make the process more general than that given on p. 660.

Solids Dissolving in Liquids

Nonpolar solids dissolve in nonpolar liquids whereas polar and ionic solids dissolve in polar liquids. For example, $I_2(s)$ readily dissolves in CCl_4 but not in H_2O; it is only slightly soluble in H_2O. In addition, polar and hydrogen-bonded sugar molecules, like sucrose, readily dissolve in water, because sugar molecules can form hydrogen bonds to water molecules. *However, network solids* like graphite, diamond and quartz sand (SiO_2) are held together by strong covalent bonds between atoms and *are not soluble in liquids, especially water. The covalent bonds in network solids are much too strong to be replaced by weaker interactions with water molecules.*

Dissolving Ionic Solids:

Ionic solids are not appreciably soluble in nonpolar liquids and their solubility in water varies greatly from compound to compound (Fig. 4.8 in the text.) *To help us understand this variance, the dissolving process can be broken down into two steps:*

• the ionic lattice is broken to give gaseous ions in an endothermic step: $MX(s) \rightarrow M^{n+}(g) + X^{n-}(g)$, and

• the gaseous ions are hydrated to give aqueous ions in an exothermic step: $M^{n+}(g) + X^{n-}(g) \xrightarrow{H_2O} M^{n+}(aq) + X^{n-}(aq)$.

If the energy required to overcome the cation-anion forces and break the lattice of the solid significantly exceeds the energy released by the ion-dipole attractions causing hydration, the $\Delta H_{solution}$ is significantly positive, and the compound is insoluble. On the other hand, if the energy released by hydra-

tion exceeds the energy required to break the lattice, the $\Delta H_{solution}$ is negative and the compound is soluble. As $\Delta H_{solution}$ becomes more and more negative, solubility increases.

The relation between $\Delta H_{solution}$ and solubility actually becomes clearer when the $\Delta H_{solution}$ values are obtained from enthalpies of formation.

$$\Delta H_{solution} = \Delta H_f^o \ MX(aq) \ - \ \Delta H_f^o \ MX(s) \qquad (14g)$$

This is because it is hard to obtain good values for both the lattice energy and hydration energy for any given salt. ✳ Try preparing a "flash card" that summarizes the impact of lattice energies and hydration energies on the solubilities of ionic solids in water and memorizing it.

Factors Affecting Solubility: Pressure and Temperature

Pressure and temperature affect the solubility of gases in liquids, but only temperature affects the solubility of solids in liquids.

Dissolving Gases in Liquids: Henry's Law

The solubility of a gas increases as the gas pressure increases. This is a statement of **Henry's law**,

$$S_g = k_H P_g \qquad (14h, 14.4)$$

where S_g is the gas solubility, P_g is the partial pressure of the gaseous solute and k_H is a constant characteristic of the solute and solvent called Henry's law constant. However, Henry's law is not applicable to gases which react chemically with the solvent.

The formation of carbonated beverages is an illustration of Henry's law. The beverage is prepared under conditions of high pressures of CO_2. When the beverage is opened, the pressure of CO_2 above the beverage is lowered and CO_2 bubbles out of the solution. In addition, Henry's law applies to underwater diving as discussed with Fig. 14.7 in the text.

Temperature Effects on Solubility: LeChatelier's Principle

Gases that dissolve to an appreciable extent in liquid solvents usually do so in exothermic processes, so $\Delta H_{solution}$ is negative. The dissolving process continues until a saturated solution is formed and the dissolving process is exactly counterbalanced by gas molecules coming out of solution with consumption of heat energy. This gives

gas + liquid solvent ⇌ saturated solution + heat energy   (14i)

If a solution of a gas in a liquid is heated, the system shifts in an attempt to absorb some of the added heat energy, so the solubility of the gas decreases with increasing temperature. This is in accord with **LeChatelier's principle**, which states that a change in any of the factors determining an equilibrium causes the system to adjust to reduce or counteract the effect of the change.

When ionic solids dissolve in liquids, the process can be endothermic or exothermic. If the process is endothermic, solubility increases with increasing temperature, because heat can be considered to be a reactant in an endothermic process. On the other hand, if the process is exothermic, solubility increases decreases with increasing temperature, as it does with gases. Thus, both trends are shown in Fig. 14.9 in the text.

The data of Fig. 14.9 also indicate the solubilities of the alkali metal chlorides increase in the order NaCl < KCl < RbCl < CsCl. Recall the discussion of the lattice energies of the alkali metal halides in Chapter 13, and note that this order is one of decreasing lattice energy with increasing cation size. *In general, the smaller the lattice energy of a salt, the greater its solubility.* However, the solubility of LiCl is an exception, and this is a reflection of the fact that *the properties of the first member of a group often differ from those of the lower members of the group.*

14.3 Colligative Properties

Properties which change between pure solvents and solutions and ideally depend only on the relative number of solute particles per solvent molecule are called the **colligative properties** of the solution.

Changes in Vapor Pressure: Raoult's Law

Water molecules can leave the surface of an aqueous solution and enter the vapor phase. However, *not as many water molecules are present at the surface of a solution as in pure water, so the vapor pressure of a solution is lower than that of pure water at the same temperature. Indeed, the vapor pressure of the solvent at the surface,* $P_{solvent}$, *is proportional to the relative number of solvent molecules in the solution, that is to the mole fraction of solvent,* $\chi_{solvent}$, *so*

$$P_{\text{solvent}} = \chi_{\text{solvent}} \cdot P^o_{\text{solvent}} \qquad (14j, 14.6)$$

where $P^o_{solvent}$ *is the vapor pressure of the pure solvent.* This relation is called **Raoult's law.**

Boiling Point Elevation

The addition of a nonvolatile solute to a liquid solvent lowers the vapor pressure in accord with Raoult's law and causes the solution to have a higher boiling point than the pure solvent, as shown in Fig. 14.12 in the text. *The greater the number of solute particles for a fixed amount of liquid, the higher the mole fraction of solute, the lower the mole fraction of the solvent and the lower the vapor pressure of the solution and the greater the boiling point elevation. The boiling point elevation is actually related to the molality of the solution by*

$$\Delta t_{bp} = K_{bp} \cdot m_{solute} \qquad (14k, 14.7)$$

where the proportionality constant K_{bp} is called the **boiling point elevation constant.** It corresponds to the elevation caused by a one molal solution and has a different value for each solvent, as shown in Table 14.4 in the text.

Using Boiling Point Elevation and Other Colligative Properties to Determine Molar Masses:

Ex. 14.6 in the text illustrates the procedure for calculating the boiling point elevation when the masses of the solute and solvent are both known and the boiling point elevation constant is known. The first step is to calculate the moles of solute to calculate the molality, m, of the solution. Conversely, *we can measure the boiling point elevation of a solution of known masses of solute and solvent to determine the molality of the solution, and use the experimental molality to determine the molar mass of the solute.* This procedure is illustrated in Ex. 14.7 in the text. *In general, the procedure is to:*

• use the change in colligative property (vapor pressure, boiling point, freezing point or osmotic pressure) to determine the concentration of the solution.

• use the definition of the concentration unit to determine the moles of solute present. In the case of boiling point elevation and freezing point depression and molality, this involves using m determined from Expression 14k or 14ℓ and the definition of molality as mol solute/kg solvent: mol solute = $m \cdot$kg solvent. In the case of osmotic pressure and molarity, this involves using the M determined from Expression 14o and the definition of molarity as moles of solute/liter of solution: mol solute = M·L of solution.

• use the moles of solute obtained from the experimental m or M and the grams of solute used to prepare the solution with the definition of moles as g/molar mass to calculate the molar mass of the solute: molar mass = g/moles.

These steps are outlined in less detail in Problem-Solving Tips and Ideas 14.1 in the text. * Try preparing a "flash card" summarizing the steps given here and memorizing them. Also practice using these steps until you are comfortable doing these types of calculations and can do them quickly.

Freezing Point Depression

The addition of a nonvolatile solute to a liquid solvent also causes the solution to have a lower freezing point than the pure solvent. The freezing point depression is related to the molality of the solution by

$$\Delta t_{fp} = K_{fp} \cdot m_{solute} \qquad (14\ell, 14.8)$$

where K_{fp} is the **freezing point depression constant**. The K_{fp} values correspond to the depression caused by one molal solutions, and are given as negative values corresponding to negative values for Δt_{fp}.

When freezing begins at the normal freezing point of the solvent, there is tendency for some molecules to leave the solid and reenter the liquid phase and for other molecules in the liquid phase to join the solid. However, the presence of solute particles at the surface of the solid inhibits the rate at which solvent molecules can join the solid. Thus, the temperature must be lowered to slow the rate of molecules leaving the solid phase and establish an equilibrium between solid and liquid: solid ⇌ liquid. This means the freezing temperature of a solution must be lower than the freezing temperature of the pure solvent. It also means that the solution does not have a sharp melting point, because the solute becomes more and more concentrated as solvent molecules enter the solid state. However, the first temperature at which solid solvent crystals appear is usually taken to be the freezing point of the solution.

Now you know why it is necessary to add antifreeze to the radiator of an automobile and to leave it in all year long. It prevents the water from freezing and expanding to give the more open structure of ice in the winter and protects against boiling over in the summer.

Colligative Properties of Solutions Containing Ions

Colligative properties only depend on the number of solute particles per solvent particle, not on what is dissolved. Thus, the changes that are caused by soluble ionic salts and acids and bases are greater than those caused by molecular solutes that do

not form ions in solution. This was discovered by Raoult in 1884 and studied in detail by Jacobus Hendricus van't Hoff in 1887. However, it was Svante Arrhenius who provided the explanation later that year: **electrolytes exist in solutions as ions.**

The effect of soluble ionic salts on colligative properties is, however, often less than predicted on the basis of complete ionization. This is primarily due to the formation of cation-anion pairs called **ion pairs.** The **van't Hoff factor, i,** gives a measure of this tendency.

$$i = \frac{\Delta t_{fp} \text{ measured}}{\Delta t_{fp} \text{ calculated}} = \frac{\Delta t_{fp} \text{ measured}}{K_{fp} \cdot m_{solute}} \qquad (14m)$$

which leads to

$$\Delta t_{fp} \text{ measured} = K_{fp} \cdot m_{solute} \cdot i \qquad (14n)$$

Expression 14h thus tells us that **the actual freezing point depression caused by an electrolyte is greater than caused by a nonelectrolyte of equal molality by the factor i. In the case of soluble salt electrolytes, i approaches the total number of moles of ions expected as the solution becomes more dilute,** as shown in Table 14.5 in the text. **In other words, ion-pairing decreases as the solution becomes more dilute and increases as the solution becomes more concentrated.** People who live in cold climates take advantage of the greater impact of electrolytes on freezing points by throwing salt on ice, rather than sugar, to lower the freezing point and melt the ice. However, adding salt to a radiator of a vehicle, rather than ethylene glycol, would accelerate the corrosion of the radiator.

Osmosis

Osmosis is the movement of solvent molecules through a semipermeable membrane from a region of lower solute concentration to a region of higher solute concentration. **Semipermeable membranes,** such as cellophane or vegetable tissue, are membranes that only allow the passage of solvent molecules. **Solvent therefore moves from the region of lower solute concentration to the region of higher solute concentration because more molecules are striking the membrane per unit time on the less concentrated side than are doing so on the more concentrated side.**

The flow of solvent between a solution and pure water continues until the pressure that is generated by the increase in solvent on the solution side exactly counterbalances the pressure of the water moving through the membrane from the pure water side. This equilibrium pressure is called the **osmotic pressure, Π,** and from experimental measurements on dilute solutions, it is known that

---

$$\Pi = MRT \qquad (14o, 14.9)$$

where Π is the osmotic pressure in atmospheres, M is the solution concentration in moles per liter; the molarity of the solution, R is the gas constant (0.0821 L·atm/mol·K) and T is the absolute temperature. Because pressures on the order of 10^{-3} atm can be easily measured, concentrations of about 10^{-4} can be measured, and *osmotic pressure is an ideal means of determining the molar masses of large molecules, many of which are of biological importance*. This type of application is illustrated in Ex. 14.10 in the text.

Osmosis is involved in the uptake of water by plants and trees, the making of pickles by placing cucumbers in brine (salt) solutions and the dehydration of bacteria cells by the application of aftershave. It is also of importance in the administration of intravenous fluids. Intravenous fluids must have the same osmotic pressure as the patient's blood, so we say it must be isosmotic or **isotonic**. If the fluid is too dilute, it is **hypotonic** and enables water to enter the blood cells and causes the cells to burst. If the fluid is too concentrated, it is **hypertonic** and enables water to leave the blood cells and causes the cells to shrivel up. A 0.16 M NaCl solution is isotonic with the cells of the body and can be used as a sterile saline solution to rehydrate a person suffering from dehydration.

14.4 Colloids

No settling of particles occurs with true solutions, such as solutions of salt or sugar, because the solute particles are in the form of ions or small molecules. In the case of suspensions, such as sand in water, the particles are large enough to be visible and settle to the bottom of the container. **Colloidal dispersions** represent a state that is intermediate between a solution and a suspension. Examples of colloids are given in Table 14.6 in the text.

Colloid particles have very high molecular masses, are large enough to disperse visible light when dispersed in a solvent and therefore make the mixture appear cloudy (Tyndall effect) and are large, but not so large that they settle out of solution. Thomas Graham, who coined the word **colloid** from the Greek word meaning "glue", also gave us the words **sol** for a dispersion of a solid substance in a fluid medium and **gel** for a dispersion that has developed a structure that prevents it from being mobile.

Types of Colloids

Colloids are classified according to the state of the dispersed phase and the dispersing medium, as noted in Table 14.6 in the text. Colloids with water as the dispersing medium can be classified as **hydrophobic** (from the Greek, meaning "water-fearing") or **hydrophilic** (from the Greek, meaning "water-loving"). *Only weak attractive forces exist between the particles of a hydrophobic colloid and water whereas the particles of a hydrophilic often contain -OH and $-NH_2$ groups on the surface that can hydrogen bond to water.*

A hydrophobic colloid is formed when a salt is formed by precipitation, because the process occurs too fast for the particles to come together to form a large crystal. Furthermore, the particles acquire a charge by absorbing additional ions of one type and are then surrounded by ions of opposite charge, so there is repulsion between the particles and they are kept from joining together to form larger crystals. This is illustrated in Fig. 14.19 in the text. However, a stable hydrophobic colloid can be made to **coagulate** by introducing ions into the dispersing medium as with the souring of milk and the formation of silt as muddy rivers reach the salt water of the seas. This is also the reason water treatment plants add $Al_2(SO_4)_3$ to water.

Proteins and starch form important hydrophilic colloids in water. However, homogenized milk is the most familiar example of a hydrophilic colloid.

Emulsions are colloidal dispersions of one liquid in another. The components of an emulsion are usually held together by the presence of an **emulsifying agent,** such as a soap or a protein, or they would quickly separate into polar and nonpolar layers. For example, lecithin from egg yolks keeps the oil and vinegar of mayonnaise together.

Surfactants

Soaps and detergents are emulsifying agents. Soap is made by heating a fat with sodium or potassium hydroxide, and the reaction that takes place is known as an ester hydrolysis reaction (11.6). *Soap has a nonpolar, hydrophobic hydrocarbon tail that is able to dissolve grease and grime and a polar, hydrophilic head that is soluble in water and keeps the soap from clinging to the grease*. The grease or oil can thus mix with water.

Substances that affect the properties of surfaces, and thus affect the interaction between two phases, are called surface-active agents or **surfactants.** *A surfactant lowers the surface tension of water.* One detergent that acts as a surfactant is biodegradable sodium lauryl benzenesulfonate; see p. 689 in the text. *Synthetic detergents commonly use sulfonate, $-SO_3^-$, groups as the polar heads instead of the carboxylate, $-CO_2^-$, groups of soaps, because the sulfonate salts formed with the Mg^{2+} and Ca^{2+} cations of hard water are more soluble than the carboxylate salts and thus do not give the scum that is obtained with soaps.*

14.5 Key Expressions

Solution concentration unit expressions:

   Molality expression:

$$\text{Molality of A } (m) = \frac{\text{moles of A}}{\text{kilograms of solvent}}$$

   Mole fraction expression:

$$\chi_A = \frac{\text{moles of A}}{\text{moles of A + moles of water}}$$

   Weight % expression:

$$\text{Wt.\% A} = \frac{\text{grams of A}}{\text{grams of solution}} \times 100\%$$

$$\text{Wt.\% A} = \frac{\text{grams of A}}{\text{grams of A + grams of water}} \times 100\%$$

   Parts per million expression:

$$\text{ppm A} = \frac{\text{milligrams of A}}{\text{L of solution}}$$

Henry's law expression:

$$S_g = k_H P_g$$

Raoult's law expression:

$$P_{\text{solvent}} = \chi_{\text{solvent}} \cdot P_{\text{solvent}}^o$$

Boiling point elevation expression:

$$\Delta t_{\text{bp}} = K_{\text{bp}} \cdot m_{\text{solute}}$$

Freezing point depression expression:

$$\Delta t_{fp} = K_{fp} \cdot m_{solute}$$

van't Hoff factor expression:

$$i = \frac{\Delta t_{fp} \text{ measured}}{\Delta t_{fp} \text{ calculated}} = \frac{\Delta t_{fp} \text{ measured}}{K_{fp} \cdot m_{solute}}$$

Freezing point depression for electrolyte solution expression:

$$\Delta t_{fp} \text{ measured} = K_{fp} \cdot m_{solute} \cdot i$$

Osmotic pressure expression:

$$\Pi = MRT$$

14.6   Hints for Answering Questions and Solving Problems

Working with molarity and molality as concentration units:

The difference between molarity and molality is a subtle, but significant difference; molarity, M, is moles of solute per liter of solution whereas molality, m, is moles of solute per kilogram of solvent. Thus, molarity is volume-based and is in terms of volume of solution, expressed in liters, whereas molality is mass-based and is in terms of mass of solvent, expressed in kilograms. Always use units with every quantity to avoid confusion when working with either of these concentration units.

Predicting solubilities of substances:

Nonpolar solutes dissolve in nonpolar solvents, and polar and ionic substances dissolve in polar solvents. The key to predicting whether a substance is nonpolar or polar is to recall that any totally symmetrical arrangement of atoms and lone pairs of electrons is nonpolar. If you need practice in predicting the polarity of substances, you should review Sections 9.5 and 9.6.

Calculating the boiling point or freezing point of a solution of a nonelectrolyte, a molecular electrolyte that does not produce ions:

Expressions 14k and 14ℓ can be used to calculate boiling point and freezing point changes, respectively. These changes can be used with the boiling point and freezing point of the pure solvent to calculate the boiling point and freezing point of the

---

solution.

$$bp_{solution} = bp_{solvent} + \Delta t_{bp}$$

$$fp_{solution} = fp_{solvent} + \Delta t_{fp}$$

Determining molar masses using colligative properties:

Follow the steps outlined on pp. 185 and 186 and used in solving Exs. 14.7 and 14.10 in the text.

CHAPTER 15

PRINCIPLES OF REACTIVITY: CHEMICAL KINETICS

Chemical kinetics involves the study of factors affecting rate of reaction and the study of the detailed step-by-step processes by which the reaction takes place, the mechanism by which the reaction takes place. On the macroscopic level, chemists are concerned about how rate depends on the concentrations of reactants, the effect of temperature and the effect of catalysts. On the submicroscopic level, chemists are concerned about the detailed manner in which atoms and molecules react. The theory which connects the macroscopic and microscopic aspects of rates of reactions is called the **collision theory of reaction rates**, because the reactants must collide to be able to react.

Chemical kinetics is a very important area of chemistry, because it is concerned with the rates of reactions that take place within us and around us. Our health and most of the products we consume are affected by the area of chemical kinetics.

15.1 Rates of Chemical Reactions

The concentrations of reactants decrease and the concentrations of products increase during a chemical reaction. The **rate of reaction** refers to the change in the concentration of a reactant or product per unit of time, Δconcentration/Δtime. Changes in concentrations can usually be measured by observing changes in properties, such as intensity of color, pressure, pH or electrical conductivity.

The rate calculated by using Δconcentration/Δtime corresponds to an **average rate** for the time interval Δt. Tables of concentration-time data or plots of concentration versus time can be used to calculate average rates during differing time intervals. These calculations show that *average rate of reaction generally decreases with time.*

Chemists are usually more interested in knowing the rate of reaction at a given instant in time, the **instantaneous rate of reaction**. *The instantaneous rate of reaction can be calculated by drawing a tangent to a concentration vs. time plot at the time of interest and taking the slope of the tangent*, as shown in Fig. 15.2 in the text.

Reactions Rates and Stoichiometry:

Perhaps you've realized that *rate of reaction can depend on which substance is being followed, because stoichiometry affects the relative rates of reaction of reactants and products.* For the reaction $N_2 + 3H_2 \rightarrow 2NH_3$, we define rate as:

$$\text{rate} = -\frac{\Delta[N_2]}{\Delta t} = -\frac{1}{3}\frac{\Delta[H_2]}{\Delta t} = \frac{1}{2}\frac{\Delta[NH_3]}{\Delta t} \tag{15a}$$

where brackets are used to represent molar concentrations (molarities) and minus signs are used with the reactants because their concentrations are decreasing and rate is expressed as a positive quantity. In general, the rate is given by

$$\text{rate} = \frac{1}{\text{stoichiometric coefficient}}(-\Delta[R]/\Delta t)$$

$$= \frac{1}{\text{stoichiometric coefficient}}(\Delta[P]/\Delta t) \tag{15b}$$

where R represents a reactant and P represents a product.

15.2 Reaction Conditions and Rate

For a chemical reaction to occur, reactant molecules must come together so atoms can be exchanged or rearranged. Thus, most reactions are carried out in the gas phase or in solution. Under these conditions, three factors affect the rate of reaction:

- *concentrations of reactants.* Magnesium metal reacts slower with 3 M HCl than it does with 6 M HCl. *Rate generally increases with increasing concentrations of reactants.*

- *temperature of reactants.* Foods cooked in water cook faster the higher the temperature of the water. *Rate generally increases with increasing temperature of the reactants.*

- *catalysts.* Catalysts are substances that accelerate chemical reactions but are not themselves transformed and thus do not appear in stoichiometric equations. Perhaps you've experienced the catalytic effect of enzymes in blood on the decomposition of hydrogen peroxide used as a disinfectant. *Rate increases in the presence of a catalyst.*

In addition, if one of the reactants is a solid, rate of reaction increases with increasing surface area. Powdered magnesium metal reacts more rapidly with a fixed amount of 3 M HCl than a single piece of magnesium of equal mass. Furthermore, this is why explosions often occur at grain storage units. *The smaller the particles of the solid, the greater the surface area available for contact with other reactants, and the faster the reaction.* ✶ Try preparing a "flash card" listing the factors affecting rate of reaction and summarizing the variation in rate that can be expected with changes in these factors and memorizing it.

15.3   Effect of Concentration on Reaction Rate

The Rate Equation

The relation between reactant concentrations and rate of reaction is given by an equation called a **rate equation**, or **rate law**. *For a general homogeneous reaction of the type*

$$aA + bB \xrightarrow{\ \ C\ \ } xX \tag{15c}$$

in which the reactants and the catalyst C are in the same phase, the rate equation has the form

$$\text{Rate} = k[A]^m[B]^n[C]^p \tag{15d}$$

where the proportionality constant k is called the **rate constant**. *The exponents m, n and p are not necessarily related to the stoichiometric coefficients and must be determined experimentally. They are usually small positive whole numbers but can also be negative numbers, fractions or zero.*

The Rate Constant k

The rate constant k is a proportionality constant that relates rate and concentration at a given temperature. It is characteristic of the reaction being studied and is therefore often called the specific reaction rate constant. It is equal to the rate of reaction when all of the concentrations in the rate equation are one molar, since one raised to any power is still one. The value of k must be determined by experiment, and once known, it can be used to calculate the rate at any set of concentrations at the same temperature.

The Order of a Reaction

The **order** with respect to a reactant is the exponent to which its concentration is raised in the rate equation, and the **total**

reaction order is the sum of the exponents of all the concentration terms. If the order for a reactant is one, it means that cutting the concentration of that reactant in half will also cut the rate in half, because $(\frac{1}{2})^1 = \frac{1}{2}$. Similarly increasing the concentration of that reactant by a factor of 4 will cause the rate to increase by a factor of 4, because $(4)^1 = 4$. On the other hand, if the order for a reactant is two, increasing the concentration of that reactant by a factor of 3 will cause the rate to increase by a factor of 9, because $(3)^2 = 9$. *If the order for a reactant is zero, it means that changes in the concentration of that reactant do not affect the rate*, because any concentration raised to the zero power is equal to 1.

Orders of reaction thus tell us how rate will change with changes in concentrations at a given temperature. Orders of reaction also give us some insight into how the reaction occurs. This point is discussed in detail in Section 15.6.

Determination of the Rate Equation

One method for determining the relation between rate and concentrations is called the **initial rates method**. The **initial rate** of reaction is the rate during the first few percent of reaction. *The initial rate can be measured by mixing the reactants and determining $-\Delta[R]/\Delta t$ or $\Delta[P]/\Delta t$ for the first 1 or 2% of the reaction,* for consumption of the first 1 or 2% of the limiting reactant. *Measuring the initial rate is convenient because the initial concentrations of the reactants are known and can be varied to determine their effect on the rate, and complications caused by the occurrence of other reactions, including the reverse of the reaction of interest, are avoided.*

The initial rates method involves varying the initial concentrations of the reactants and systematically determining the effect on the initial rate of reaction. If the initial concentration of a particular reactant is doubled while holding all other initial concentrations the same and the initial rate doubles, the reaction must be first order with respect to that reactant, because $(2)^x = 2$ and x must be one. On the other hand, if the initial rate increases by a factor of 4, the reaction must be second order with respect to that reactant, because $(2)^x = 4$ and x must be two. *The general relation between concentration and rate for a reactant with an order x is given mathematically by*

$$\left(\frac{\text{Initial conc 2}}{\text{Initial conc 1}}\right)^x = \left(\frac{\text{Initial rate 2}}{\text{Initial rate 1}}\right) \tag{15e}$$

The general procedure for using initial rates to determine the rate equation for the general reaction aA + bB $\rightarrow$ cC having the general rate equation Rate = $k[A]^m[B]^n$ is to:

• vary the initial [A] while holding [B] constant and determining the effect of [A] on the initial rate to determine the value of m. [A] is usually varied by several factors, so it could be cut in half, doubled, tripled, etc.

• vary the initial [B] while holding [A] constant and determining the effect of [B] on the initial rate to determine the value of n. [B] is usually varied by several factors, so it could be cut in half, doubled, tripled, etc.

• solve the rate equation for k and use the data for each experiment to calculate a value of k for each experiment and then calculate an average value of k. That is, use

$$k = \frac{Rate}{[A]^m[B]^n} \qquad (15f)$$

with the experimental values of m and n and the concentration and rate data for each experiment.

• rewrite the general rate equation using the average value of k for the temperature of the study and the experimental values of m and n.

The rate equation thus obtained can then be used to calculate the reaction rate at any set of concentrations at the temperature used for the initial rate study.

Units for k:

The units for k vary with the order of reaction, as suggested by Expression 15f. The units used for rate throughout this chapter are mol/L·time, so the units for k vary with the values of m and n and therefore with the overall order of reaction given by $m + n$. For time in seconds, we obtain:

Overall Order	Units of k		
0	mol/L·s	or	$mol \cdot L^{-1} \cdot s^{-1}$
1	1/s	or	s^{-1}
2	L/mol·s	or	$L \cdot mol^{-1}$
3	$L^2/mol^2 \cdot s$	or	$L^2 \cdot mol^{-2} \cdot s^{-1}$
n	$L^{n-1}/mol^{n-1} \cdot s$	or	$L^{n-1} \cdot mol^{-(n-1)} \cdot s^{-1}$

In other words, the units for k must be such that when multiplied by concentrations raised to their appropriate powers, the overall product of the units must be the units for rate, namely mol/L·time.

15.4   Relationships Between Concentration and Time

The tools of calculus can be used to derive equations, called integrated rate equations, which relate concentration to time and can be used to calculate:

- the value of k knowing the initial concentration and the concentration after some amount of time t has elapsed,

- the concentration remaining at some time t knowing the initial concentration and the value of k,

- the initial concentration knowing the concentration remaining at some time t and the value of k, or

- the time required to get from an initial concentration to a given concentration knowing the value of k.

First-Order Reactions

For the first-order reaction $R \rightarrow$ products, the (differential) rate equation is

$$\text{Rate} = -(\Delta[R]/\Delta t) = k[R] \tag{15g}$$

and the integrated rate equation is

$$\ln \frac{[R]_t}{[R]_o} = -kt \tag{15h, 15.1}$$

Here $[R]_o$ is the concentration at time $t = 0$, the initial concentration. (For first-order reactions this does not need to correspond to the actual beginning of the experiment, rather it can correspond to when the instrument readings are started.) **$[R]_t$ is the concentration at time t, ($[R]_t/[R]_o$) is the fraction of reactant R remaining at time t and k is expressed in 1/time.**

Exs. 15.3 and 15.4 in the text illustrate the use of Expression 15i to solve for time and $[R]_t$, respectively. Appendix A in the text gives further discussion of logarithms, their use, and their relation to the proper number of significant figures. *In general, numbers to the left of the decimal in a logarithmic quantity are related to the power of 10 for the number, and only those digits beyond the decimal are considered to be significant figures*, so 3.40 has two significant figures as a logarithmic quantity, the 4 and the 0.

Second-Order Reactions

For the second-order reaction $R \rightarrow$ products or $R_1 + R_2 \rightarrow$ products with equal initial concentrations of R_1 and R_2, the (differential) rate equation is

$$\text{Rate} = -(\Delta[R]/\Delta t) = k[R]^2 \qquad (15i)$$

and the integrated rate equation is

$$\frac{1}{[R]_t} - \frac{1}{[R]_o} = kt \qquad (15j, 15.2)$$

where k is expressed in L/mol·time. Ex. 15.5 in the text illustrates the use of Expression 15k to solve for time, and Exercise 15.7 in the text requires solving for $[R]_t$.

Zero-Order Reactions

For the zero-order reaction $R \rightarrow$ products, the (differential) rate equation is

$$\text{Rate} = -(\Delta[R]/\Delta t) = k[R]^0 = k \qquad (15k)$$

and the integrated rate equation is

$$[R]_o - [R]_t = kt \qquad (15\ell, 15.3)$$

where k is expressed in mol/L·time. ✳ Try preparing a separate "flash card" for each order giving the (differential) rate equation and integrated rate equation for the order and memorizing them. Also try using the integrated rate equations until you are comfortable using them and can use them quickly.

Graphical Methods for Determining Reaction Order and Rate Constant:

The integrated rate equations for zero, first and second-order reactions can be rearranged to fit the equation for a straight line: $y = a + bx$, where a is the intercept value of y when x equals zero and b is the slope of the line (See the margin figure on p. 714 in the text.) This gives

- *for zero order:* $\qquad [R]_t = [R]_o - kt \qquad (15m)$

   This means the reaction is zero order, if a plot of $[R]_t$ versus time gives a straight line. The intercept of the line will be $[R]_o$, and the slope will be $-k$. See Fig. 15.7 in the text.

- **for first order:** $\ln[R]_t = \ln[R]_o - kt$ (15n)

  This means the reaction is first order, if a plot of $\ln[R]_t$ versus time gives a straight line. The intercept of the line will be $\ln[R]_o$, and the slope will be $-k$. See Fig. 15.8 in the text.

- **for second order:** $\dfrac{1}{[R]_t} = \dfrac{1}{[R]_o} + kt$ (15o)

  This means the reaction is second order, if a plot of $1/[R]_t$ versus time gives a straight line. The intercept of the line will be $1/[R]_o$, and the slope will be k. See Fig. 15.9 in the text.

In summary, a straight line is only obtained when

- **$[R]_t$ is plotted versus time for a zero-order reaction.** All other order reactions give curved lines for plots of $[R]_t$ versus time.

- **$\ln[R]_t$ is plotted versus time for a first-order reaction.** All other order reactions give curved lines for plots of $\ln[R]_t$ versus time.

- **$1/[R]_t$ is plotted versus time for a second-order reaction.** All other order reactions give curved lines for plots of $1/[R]_t$ versus time.

Thus, to determine the order of reaction, chemists can plot the experimental concentration-time data in different ways until a straight-line plot is achieved. The characteristics of these plots are summarized quite nicely in Table 15.2 in the text.
✶ Try preparing a "flash card" summarizing this information and memorizing it or try preparing a separate "flash card" for each reaction order and memorizing them.

Half-Life and First-Order Reactions

The **half-life**, $t_{1/2}$, of a reaction is the time that is required for the concentration of a reactant to decrease to one half its initial value. For a reaction that is first-order in R and first-order overall, $t_{1/2}$ **is the time when**

$$[R]_t = 1/2\,[R]_o \quad \text{or} \quad [R]_t/[R]_o = 1/2$$

Substituting this into the first-order concentration-time

expression, Expression 15h, gives

$$\ln(1/2) = -kt_{1/2}$$

which leads to the **first order half-life expression**

$$t_{1/2} = \frac{0.693}{k} \qquad (15p)$$

Notice that $t_{1/2}$ is independent of concentration for first order reactions. This means that if it takes 20 min to go from 1.0 M to 0.5 M for a reactant at a given temperature, it will also take 20 min to go from 0.8 M to 0.4 M for the same reactant at the same temperature. *It also means that each successive half-life takes the same amount of time for a first-order reaction. Furthermore, the fraction of reactant remaining after n half-lives is always equal to $(1/2)^n$, as it is with every order.* The last two concepts are discussed on p. 717 and illustrated graphically in Fig. 15.10 on p. 718 in the text. ✴ Try preparing a "flash card" containing Expression 15p and memorizing it. Also try preparing a "flash card" containing the equation for calculating the fraction of reactant remaining after n half-lives, as well as the actual fraction remaining after the first five half-lives and memorizing it.

15.5 A Microscopic View of Reactions

The goal of this section is to use the kinetic molecular theory of matter to provide microscopic explanations for the macroscopic factors affecting rates of reactions. Our explanations will be based on the **collision theory** of reaction rates, which assumes molecules must collide with one another to react.

Effect of Nature of the Reactants on Reaction Rate-Activation Energy

Every chemical reaction has an energy barrier that must be surmounted if the colliding molecules are to react. The energy barrier is due to the fact that the colliding molecules must possess sufficient kinetic energy to overcome the electron cloud-electron cloud repulsions between them and transfer sufficient kinetic energy to begin to break bonds in them as reacting molecules. The heights of these barriers vary from almost zero to hundreds of kJ/mol. *For similar reactions at the same temperature, the higher the barrier, the slower the reaction.* The height of the barrier is called the **activation energy** for the reaction.

At any given temperature there is a distribution of kinetic energies, as illustrated in Fig. 15.12 in the text, **and only a fraction of the molecules possess sufficient kinetic energy to surmount the barrier, to overcome the activation energy. The lower the activation energy, the higher the fraction of molecules with sufficient kinetic energy to overcome the activation energy barrier at any given temperature. Thus, at any given temperature, reactions with low activation energies tend to proceed faster than those with high activation energies.**

Effect of Reactant Concentrations on Reaction Rate

The higher the concentrations of the reactants, the higher the number of molecules with sufficient kinetic energy to react at any given temperature, and the faster the reaction.

Effect of Molecular Orientations on Reaction Rate

Only a fraction of the collisions between molecules with sufficient kinetic energy to react involve molecules coming together with proper orientation to react. This so-called **steric factor** is particularly important for atom transfer reactions, such as $NO(g) + O_3(g) \rightarrow NO_2(g) + O_2(g)$. In this reaction, an O atom transfers from O_3 to NO, so the N atom end of the NO molecule must collide with an O atom from O_3 for this to occur. *In general, the more difficult it is to achieve the proper orientation for reaction, the slower the reaction.*

Effect of Temperature on Reaction Rate: The Arrhenius Equation

Reaction rate generally increases with increasing temperature, because the higher the temperature, the greater the fraction of molecules with sufficient kinetic energy to overcome the activation energy barrier. See Fig. 15.12 in the text.

Thus, reaction rate depends on the nature of the reactants and the resulting impact on the activation energy barrier, the fraction of collisions with proper orientation and the temperature, which affects the number of collisions per unit of time and more importantly the kinetic energies of the colliding particles. These factors are brought together in the Arrhenius equation:

$$k = A e^{-Ea/RT}$$

(15q,15.5)

where

- A, the **frequency factor,** is equal to the frequency of collisions with proper orientation when the reactant concentrations = 1 M,

- $e^{-Ea/RT}$ is the fraction of molecules with KE $\geq E_a$, and

- R is the gas constant, 8.314×10^{-3} kJ/mol·K.

This equation can be used to (1) calculate the value of the activation energy E_a from the temperature dependence of the rate constant and (2) calculate the rate constant for a given temperature when the activation energy and A factor are known.

If we take the natural logarithm of Expression 15r and rearrange the result, **we obtain the following equation for a straight line relating ln k to (1/T):**

$$\ln k = \ln A - \frac{E_a}{R}\left(\frac{1}{T}\right) \qquad (15r, 15.6)$$

This means we can study k at several temperatures and plot ln k versus 1/T to determine the value of E_a, because the slope of the downward-sloping line is equal to $-E_a/R$ and the intercept is equal to ln A. There are several ways to rationalize the downward slope of such plots: (1) k increases with increasing temperature because rate increases with increasing temperature and the higher the value of k, the higher the value of ln k and simultaneously, the greater the value of T, the lower the value of 1/T. (2) The slope of such a plot is equal to $-E_a/R$, and the activation energy for a reaction represents an energy barrier to reaction, a positive quantity, so the slope must be negative to give a positive value for E_a. Ex. 15.7 in the text illustrates the use of Expression 15s for graphical determination of E_a.

E_a can also be determined algebraically. Knowing k at two different temperatures enables us to apply Expression 15r to both temperatures to obtain

$$\ln \frac{k_1}{k_2} = -\frac{E_a}{R}\left[\frac{1}{T_2} - \frac{1}{T_1}\right] \qquad (15s, 15.7)$$

which can be solved for E_a. However, this method involves taking the slope a line having just two data points, either of which could involve substantial experimental error, and isn't as likely to give an accurate value for E_a as a graphical plot using more data points. **A far more important application of Expression 15t involves calculating k_2 at some temperature T_2 when k_1, T_1 and E_a are known or calculating the temperature T_2 that is required for k_2 to be a given value when k_1, T_1 and E_a are known.** For example, we might wish to know the temperature at which a reaction will proceed 10 times faster than it does at 25°C.

15.6 Reaction Mechanisms

One of the more important reasons for studying rates of reactions is to learn something about the **reaction mechanism**, the sequence of bond-breaking and bond-making steps that occurs during the conversion of reactants to products. *All chemical equations are represented by a single balanced equation showing the overall stoichiometry of the reaction but very few reactions occur in a single step. Indeed, it is unlikely that more than two particles can collide with sufficient kinetic energy and proper orientation, unless the reaction is in aqueous solution and one of the reactants is H^+, OH^- or H_2O. Knowing the rate equation for a reaction, chemists can make an educated guess about the mechanism.*

Each of the steps in a reaction sequence, a reaction mechanism, is called an **elementary step.** *Each step has its own activation energy barrier E_a and rate constant k, and the steps must add up to give the balanced equation for the overall reaction. The set of steps that satisfactorily account for the kinetic and stoichiometric properties of a chemical reaction constitutes a reaction mechanism. In some cases, more than one mechanism can meet these criterion, and it is necessary to decide between them on the basis of other evidence whenever possible.*

Molecularity of Elementary Steps

Elementary steps are classified by the number of reactant particles that come together in a step. This number is called its **molecularity.** The order of a reaction can be a fractional number, but the molecularity is always a whole positive number.

- An elementary step involving just one molecule is a **unimolecular process**. The form of a unimolecular elementary step is A → products.

- An elementary step involving two molecules is a **bimolecular process.** The form of a bimolecular elementary step can be 2A → products or A + B → products.

- An elementary step involving three molecules is a **termolecular process.** The form of a termolecular elementary step can be 3A → products, 2A + B → products or A + B + C → products. *However, the simultaneous collision of three particles with sufficient kinetic energy for reaction and proper orientation for reaction is unlikely, unless one of the species is a solvent species, as noted above, or the* *reaction occurs in the gas phase and a species serves to carry away the energy that is produced as two active reactants react in an exothermic manner*. For example, the role of N_2 in the termolecular reaction that produces ozone, O_3, in the upper atmosphere,

$$O(g) + O_2(g) + N_2(g) \rightarrow O_3(g) + \text{energetic } N_2(g)$$

is to carry away excess energy which might otherwise cause the ozone to decompose.

Rate Equations for Elementary Steps

The rate equation for an elementary step can be written using the stoichiometry of that step. For example, the rate equations for the two elementary steps in the mechanism for the decomposition of ozone

Elementary step 1:	$O_3(g) \rightarrow O_2(g) + O(g)$	(15t)
Elementary step 2:	$\underline{O_3(g) + O(g) \rightarrow 2O_2(g)}$	(15u)
Overall reaction:	$2 O_3(g) \rightarrow 3 O_2(g)$	

are

Elementary step 1:	Rate = $k[O_3]$	(15v)
Elementary step 2:	Rate = $k'[O_3][O]$	(15w)

and the two rate constants k and k' are not expected to have the same value. Other examples of writing rate equations for elementary steps are given on p. 729 in the text. *Remember, however, that it is not possible to look at the stoichiometry of an overall equation and write a rate law for the reaction. The overall rate law can only be determined experimentally.*

Molecularity and Reaction Order

The molecularity of an elementary step and its order are the same. The relation between molecularity and order, however, is emphatically not true for the overall order reaction. Thus, a second order elementary step is a bimolecular step whereas a second order overall reaction might or might not occur in a single bimolecular step. The relation between an experimentally determined overall order of reaction and a possible mechanism or sequence of elementary steps is explored in the next section.

Reaction Mechanism and Rate Equations

Suppose you are part of an assembly line preparing ham sandwiches. Surely you realize that ham sandwiches cannot be produced any faster than the slowest step in the process. The same is true of chemical reactions. ***The products of a reaction can never be produced faster than the rate of the slowest step. The rate of the overall reaction is limited by and is exactly equal to the combined rates of all the elementary steps up through the slowest step in the mechanism.*** The slowest elementary step of a sequence is therefore called the **rate-determining step** of the mechanism.

Experiment shows that the reaction of nitrogen dioxide with fluorine, for example,

$$2 \ NO_2(g) + F_2(g) \rightarrow 2 \ FNO_2 \qquad (15x)$$

has the experimental second-order rate equation

$$Rate = k[NO_2][F_2] \qquad (15y)$$

This immediately rules out the possibility of the overall reaction occurring in a single step, because the rate equation for a single step reaction, according to the overall equation, would be

$$Rate = k[NO_2]^2[F_2] \qquad (15z)$$

and this does not agree with the experimental second-order rate equation. Thus, there must be at least two steps in the mechanism, and the rate-determining elementary step is likely to involve NO_2 and F_2 reacting in a 1:1 ratio. The simplest possibility is:

Elementary step 1:slow $NO_2(g) + F_2(g) \ -k_1->\ FNO_2(g) + F(g)$
Elementary step 2:fast $\underline{NO_2(g) + F(g) \ \ \ -k_2->\ FNO_2(g)}$
Overall: $2 \ NO_2(g) + F_2(g) \longrightarrow 2 \ FNO_2(g)$

Since the first elementary step is assumed to be the slow step, the rate-determining step, in the mechanism, the rate equation would be:

$$Rate = k_1[NO_2][F_2] \qquad (15aa)$$

in accord with the experimental rate equation with $k_{expt1} = k_1$.

The F atom formed in the first step of the NO_2/F_2 reaction is called an intermediate. A **reaction intermediate** is formed in one step of a reaction sequence and then consumed in equal amounts in a subsequent step or steps. ***Hence, an intermediate does not appear in the overall stoichiometric equation.*** Indeed, ***reaction intermediates usually have fleeting lifetimes and might or might not exist long enough to be observed.***

Mechanisms involving slow first steps represent only one type of the many possible reaction mechanisms. However, this is the only type discussed in this section. The relation between rate equations and reaction mechanisms is summarized in Problem Solving Tips and Ideas 15.2 in the text.

15.7 Catalysts and Reaction Rate

Catalysts, you recall, are substances which accelerate the rates of chemical reactions but are not themselves transformed and thus do not appear in stoichiometric equations. A **homogeneous catalyst** exists in the same phase as the reactants. *It is consumed in the rate determining step or in a step preceding the rate determining step, and then reformed in equal amounts in a step or steps following the rate determining step. Because it is consumed in the rate determining step or in a step preceding the rate determining step, its concentration necessarily appears in the rate equation for the reaction. However, because it is consumed and then reformed in equal amounts in the sequence of steps it does not appear in the overall stoichiometric equation for the reaction.* See the rate equations for the decomposition of hydrogen peroxide in the presence of iodide anions and the isomerization of cis-2-butene in the presence of iodine molecules given on pages 705 and 734 in the text, respectively.

A catalyst operates by altering the mechanism of the reaction, so the activation energy barrier is lowered. At any given temperature, more molecules possess sufficient kinetic energy to overcome the activation energy barrier, and the reaction occurs more rapidly. The lowering of the activation energy is illustrated in the potential energy diagrams given in Fig. 15.17 in the text for both the uncatalyzed and iodine catalyzed isomerization reactions.

Catalysis in Industry and the Environment

It has been estimated that 90% of all manufactured items use catalysts at some stage of production and that more than a trillion dollars worth of goods is manufactured with the aid of manmade catalysts. The conversion of crude oil to gasoline is just one of the many important industrial processes which use catalysts. In addition, the interview with James Cusumano that appears on p. 155 in the text describes the importance of catalysts in solving environmental problems.

Industrial processes commonly use **heterogeneous catalysts**, which are present in a different phase than the reactants being catalyzed. *Heterogeneous catalysts are usually solids which are not significantly altered during the course of a reaction and*

can be easily separated from the products and any leftover reactants. The processes for making nitric acid, a commercially important chemical, and for removing pollutants from exhaust gases of vehicles using catalytic mufflers use heterogeneous catalysts, as described in the text.

Reaction occurs on the surface of a heterogeneous catalyst. It is thought that the interaction between the catalyst and the reactant weakens the bonds in the reactants and thus lowers the activation energy for the process. Thus, more molecules possess sufficient kinetic energy to overcome the activation energy barrier, and the reaction proceeds more rapidly at the given temperature. In the case of catalytic mufflers, this even allows the flameless combustion of any unburned hydrocarbons to occur.

15.8 Key Expressions

Reaction rate and stoichiometry expression:

$$\text{rate} = \frac{1}{\text{stoichiometric coefficient}}(-\Delta[R]/\Delta t)$$

$$= \frac{1}{\text{stoichiometric coefficient}}(\Delta[P]/\Delta t)$$

General rate equation expression for the reaction of A and B:

For a homogeneous reaction in the absence of any catalyst,

$$\text{Rate} = k[A]^m[B]^n$$

but in the presence of a homogeneous catalyst C,

$$\text{Rate} = k[A]^m[B]^n[C]^p$$

General initial rates expression for determining the order of a reactant while holding the concentrations of all other reactants constant:

$$\left(\frac{\text{Initial conc 2}}{\text{Initial conc 1}}\right)^x = \left(\frac{\text{Initial rate 2}}{\text{Initial rate 1}}\right)$$

Differential and integrated rate expressions:

For zero-order reactions:

$$\text{Rate} = -(\Delta[R]/\Delta t) = k[R]^0 = k \quad \text{and}$$

$$[R]_o - [R]_t = kt$$

For first-order reactions:

$$\text{Rate} = -(\Delta[R]/\Delta t) = k[R] \quad \text{and}$$

$$\ln \frac{[R]_t}{[R]_o} = -kt$$

For second-order reactions:

$$\text{Rate} = -(\Delta[R]/\Delta t) = k[R]^2 \quad \text{and}$$

$$\frac{1}{[R]_t} - \frac{1}{[R]_o} = kt$$

Graphical order testing expressions:

For zero-order: $\quad [R]_t = [R]_o - kt$

For first-order: $\quad \ln[R]_t = \ln[R]_o - kt$

For second-order: $\quad \dfrac{1}{[R]_t} = \dfrac{1}{[R]_o} + kt$

First-order half-life expression:

$$t_{1/2} = \frac{0.693}{k}$$

Arrhenius expression:

$$k = Ae^{-Ea/RT}$$

"Two-point" rate constant versus temperature expression:

$$\ln \frac{k_1}{k_2} = -\frac{E_a}{R}\left[\frac{1}{T_2} - \frac{1}{T_1}\right]$$

15.9 Hints for Answering Questions and Solving Problems

Correlating rates with stoichiometry:

Consider the following general equation: aA + bB → cC. If the rate of disappearance of A is known and the rate of disappearance of B or rate of appearance of C is requested, the desired quantity can be calculated by doing a mole-to-mole conversion using the coefficients of the balanced equation. For example, if the rate of disappearance of A is known to be 0.20 mol A/L·hr, the rate of disappearance of B can be calculated by using

$$0.20 \frac{\text{mol A}}{\text{L·hr}} \cdot \frac{\text{b mol B}}{\text{a mol A}}$$

Writing rate equations:

Rate equations for elementary steps can always be written by examining the stoichiometry of the elementary steps whereas rate equations can never be written for overall reactions by examining the stoichiometry of overall equations. The reason for this lies in the fact that the orders for an overall reaction depend only on the stoichiometry of the rate determining step and any steps that come before it whereas the stoichiometric coefficients of the overall reaction depends on the sum of all of the steps of the mechanism, including any steps which follow the rate determining step. Consider the mechanism given on p. 206 for the reaction of NO_2 with F_2. The first elementary step is believed to be the rate determining step and the second step is believed to be fast. The orders for the first elementary step and therefore the reaction depend only on the stoichiometric coefficients of that step whereas the stoichiometric coefficients for the overall reaction depend on the stoichiometric coefficients for both elementary steps. Thus, **overall orders of reaction must always be determined experimentally.**

Identifying catalysts and intermediates in mechanisms:

A homogeneous <u>catalyst</u> is <u>consumed</u> in one step and <u>then reformed</u> in the same amount in a subsequent step or steps, so it never appears in the overall equation for the reaction. On the other hand, an <u>intermediate</u> is <u>formed</u> in one step and <u>then consumed</u> in the same amount in a subsequent step or steps, so it also never appears in the overall equation for the reaction.

CHAPTER 16

PRINCIPLES OF REACTIVITY:
CHEMICAL EQUILIBRIA

Chemical equilibria involves the study of extent of reaction, that is, the study of the extent of conversion of reactants to products at a given temperature. In this chapter, we will explore the characteristics of the state of chemical equilibrium and the effects of changes in factors affecting the state of chemical equilibrium. We will also consider the relation between chemical equilibrium, kinetics and reaction mechanisms.

Chemical equilibria is another very important area of chemistry, because it affects the extent of reactions that take place within us and around us. In fact, chemical equilibria is so important that it is primary topic of Chapters 16-18. We shall see that it affects such diverse areas as our blood buffer system, our dental health, the production of ammonia for use as a fertilizer, the formation of limestone caves and many other environmental issues.

16.1 The Nature of the Equilibrium State

The concept of a dynamic equilibria was introduced in Chapter 13 in conjunction with phase equilibria. Recall the case of a volatile liquid occupying a closed container. Molecules at the surface of the liquid having sufficient kinetic energy to overcome the attractive forces of their neighboring molecules escape and enter the vapor phase. However, some of the molecules in the vapor phase reenter the liquid phase, and eventually the point is reached at which liquid⇌vapor occurs at equal rates. *We describe this state as a state of dynamic equilibrium, because the processes continue at equal rates, but there is no net change in the amount of molecules in the liquid and vapor phases. This same concept applies to chemical reactions.*

In principle, all chemical reactions are reversible. Eventually a state of dynamic equilibrium is reached in which the forward and reverse reactions continue to occur, but they take place at equal rates. Indeed, *it can be shown that the nature of the equilibrium state for a specific chemical reaction is the same at any given temperature, no matter the direction of the approach. This means we can start with either reactants or products and expect to find the same distribution of reactants and products at equilibrium, as long as we are using the same*

amount of material. See the discussion of the acetate ion/acetic acid equilibria given on pp. 752 and 753 in the text for a specific example of this.

16.2 The Equilibrium Constant

An **equilibrium constant expression** relates the concentrations of reactants and products at equilibrium at a given temperature to a numerical constant. *For the general reaction*

$$aA + bB \rightleftarrows cC + dD \tag{16a}$$

the equilibrium concentrations of reactants and products are related by the equilibrium constant expression

$$\text{Equilibrium constant} = K = \frac{[C]^c[D]^d}{[A]^a[B]^b} \tag{16b, 16.1}$$

Product concentrations are always placed in the numerator, and reactant concentrations are always placed in the denominator. In addition, the concentration of each substance is raised to the power of its stoichiometric coefficient in the balanced equation. The value of K depends on the particular reaction and on the temperature and tells us about the extent of reaction that has occurred when equilibrium has been achieved.

Writing Equilibrium Concentration Expressions

Reactions Involving Solids and Water:

The molar concentration of a solid is fixed by its density and molar mass

$$\frac{g}{mL} \cdot \frac{10^3 \text{ mL}}{1 \text{ L}} \cdot \frac{1 \text{ mole}}{M \text{ g}} = \frac{\text{moles}}{L} \tag{16c}$$

and is not changed by either the reaction of some solid or the addition of some solid. Furthermore, the equilibrium concentrations of other reactants and products are not changed by the amount of solid present in a reaction, so long as some solid is present at equilibrium. Therefore, chemists do not include the concentrations of any solids in an equilibrium constant expression.

You can think of the concentrations of solids as already being incorporated in the value of K for the reaction. For example, for

$$\tfrac{1}{8}S(s) + O_2(g) \rightleftharpoons SO_2(g) \qquad (16d)$$

$$K' = \frac{[SO_2(g)]}{[S(s)]^{\frac{1}{8}}[O_2(g)]} \qquad (16e)$$

so,

$$K' \cdot [S(s)]^{\frac{1}{8}} = K = \frac{[SO_2(g)]}{[O_2(g)]} \qquad (16f)$$

The molar concentration of water is also fixed by its density and molar mass and is 55.6 M at room temperature:

$$\frac{1.00 \text{ g}}{\text{mL}} \cdot \frac{10^3 \text{ mL}}{\text{L}} \cdot \frac{1 \text{ mole}}{18.0 \text{ g}} = 55.6 \frac{\text{mol}}{\text{L}} \qquad (16g)$$

For this reason, the concentration of water is not changed by chemical reactions occurring in aqueous solutions and is therefore not included in equilibrium constant expressions for reactions occurring in aqueous solutions. You can also think of the concentration of water as being already incorporated in the value of K for reactions occurring in aqueous solutions, such as those shown for the weak base NH_3 and the weak acid HCO_2H on pages 755 and 756 in the text.

Expressing Concentrations: K_c and K_p

The concentrations in the equilibrium constant expression are usually given in moles per liter (M), as implied by the use of brackets thus far, so the equilibrium constant symbol K usually is given a subscript c for concentration, as in K_c. However, equilibrium constant expressions for reactions involving gases can also be written in terms of the partial pressures of the gaseous reactants and products, and K can be given a subscript p for pressure, as in K_p. The relation between K_c and K_p for a given reaction is explored in the box "A Closer Look:Equilibrium Expressions for Gases" on p. 755 in the text.

Manipulating Equilibrium Expressions

Chemical equations can be multiplied by factors, reversed or even added together to obtain an equation for a net process.

• *When the stoichiometric coefficients of a balanced equation are multiplied by a factor, the equilibrium constant for the new equation, K_2, is equal to the old equilibrium constant, K_1, raised to the power of the multiplication factor.* Thus, when an equation is multiplied by a factor of 2, $K_2 = (K_1)^2$.

• *When an equation is reversed, it corresponds to multiplying the equation by -1, so $K_2 = (K_1)^{-1} = 1/K_1$. The equilibrium constants for an equation and its reverse are always the reciprocal of each other, because the identity of the reactants and products has been inverted.*

• *When two or more equations are added together to obtain the equation for a net process, the equilibrium constant for the net equation is equal to the product of the equilibrium constants for the added equations.* Thus, when one equation is added to two times a second equation to obtain a third equation, $K_3 = K_1 \cdot (K_2)^2$.

Information concerning the writing and manipulating of equilibrium constant expressions is given in Problem-Solving Tips and Ideas 16.1 in the text. ✻ Try preparing a "flash card" summarizing the information of this section thus far and the information of Problem-Solving Tips and Ideas 16.1 and memorizing it.

The Meaning of the Equilibrium Constant

The value of the equilibrium constant indicates whether a reaction is reactant or product-favored. A large value of K (K>>1) means the ratio of product concentrations over reactant concentrations, each raised to the power of their stoichiometric coefficient, is large, and the reaction is product-favored. Conversely, a value of K <<1 means the concentrations of the reactants are large compared to those of the products, and the reaction is reactant favored. The actual value of K can be used with the initial concentrations and the stoichiometry of the reaction to calculate the concentrations of reactants and products at equilibrium. This use of K is illustrated in Section 16.5. ✻ Try preparing a "flash card" summarizing the meaning of the value of K and memorizing it.

16.3 The Reaction Quotient, Q

A general expression pertaining to the concentrations of reactants and products at any point in a reaction at a given temperature is given by the **reaction quotient expression**. *For the general reaction*

$$aA + bB \rightleftharpoons cC + dD \tag{16h}$$

the reaction quotient expression is:

$$\text{Reaction quotient} = Q_C = \frac{[C]^c[D]^d}{[A]^a[B]^b} \qquad (16i)$$

Thus the form of a reaction quotient expression is identical to the corresponding equilibrium constant expression, but Q_C differs from K_C in that the concentrations used in the reaction quotient expression are not necessarily equilibrium concentrations. In other words, K_C is merely a special value of Q_C pertaining to equilibrium conditions.

The Relationship Between the Reaction Quotient Q_C and the Equilibrium Constant K_C

If

- *$Q < K$, the ratio of product concentrations over reactant concentrations is too small, and reactants must be converted to products until $Q = K$.*

- *$Q = K$, the system is at equilibrium.*

- *$Q > K$, the ratio of product concentrations over reactant concentrations is too large, and products must converted to reactants until $Q = K$.*

16.4 Calculating an Equilibrium Constant

When the equilibrium concentrations of all the reactants and products are known, calculating K_C involves simply substituting the data into the reaction quotient expression, as in Ex. 16.3 in the text. However, the more common case involves knowing the initial concentrations of the reactants and the equilibrium concentration of just one reactant or product that has been measured experimentally. Equilibrium concentrations for the rest of the reactants and products must then be calculated by using the change in the concentration for that one substance with the stoichiometry of the reaction to determine the changes for the other substances leading to equilibrium. Finally, these changes are used with the initial concentrations to calculate the equilibrium concentrations and the value of the equilibrium constant.

Equilibrium tables called **ICE tables** can be used to summarize the data used for calculating K_C. *The I in ICE stands for initial, the C for change and the E for equilibrium. The entries for I, C and E can be in terms of moles, but molar concentrations are always required for substituting into the reaction quotient expression and calculating K_C. In the case of reactions* *involving only gaseous reactants and products, the entries can be in terms of partial pressures, which can be used to calculate K_p.* These principles are illustrated in Exs. 16.4-6 in the text. *Notice how the initial concentration and equilibrium concentration of the one substance for which such information is known is used to determine the change for that substance, and how this change is then used with the stoichiometric coefficients of the balanced equation to do mole-to-mole conversions to calculate the changes for all other reactants and products. Also notice the importance of expressing changes in terms of just one unknown, x.* * Try preparing a "flash card" summarizing the procedures for conducting these calculations and practice using them until you are comfortable doing them and can do them quickly.

16.5 Using Equilibrium Constants in Calculations

The value of K_c can be used with the initial concentrations to calculate the concentrations present at equilibrium. The procedure involves:

- constructing an appropriate ICE table that expresses the changes in terms of just one unknown, x,

- substituting the expressions for the equilibrium concentrations into the appropriate equilibrium constant expression,

- solving for x and

- substituting the value of x into the expressions for the equilibrium concentrations to calculate the equilibrium concentrations of all reactants and products.

The value of K_p can similarly be used with the initial pressures to calculate the partial pressures present at equilibrium for gaseous reactions.

The steps outlined above are used in solving for equilibrium concentrations in Exs. 16.7 and 16.8 in the text, which illustrate three common ways of solving such problems:

- Ex. 16.7 illustrates the **perfect-square method**, which involves taking the square root of both sides to simplify the mathematics whenever the equilibrium constant expression is a perfect square.

- Ex. 16.8 illustrates the **quadratic method**, which involves using the quadratic equation to solve for the value of x and gives two roots as answers, one of which usually leads to negative concentrations and is physically meaningless.

• Ex. 16.8 also illustrates the **approximation method**, which involves assuming x is small compared to the initial concentrations of reactants and products, when the initial concentrations are greater than zero, and thus avoids the use of the quadratic or higher equation to solve for x. *This method can be used whenever the calculated value of x turns out to be less than 5% of the smallest initial concentration, as determined from*

$$\frac{x}{\text{smallest initial concentration}} \cdot 100\% \qquad (16j)$$

This magnitude of error is justifiable, because equilibrium constants themselves usually are not known to better than ±5%.

The latter method is the most common method and will be used extensively, especially in Chapters 17 and 18.

16.6 Disturbing a Chemical Equilibrium

LeChatelier's Principle

There are three common ways in which a chemical reaction system at equilibrium can be disturbed: (1) by a change in temperature, (2) by a change in the concentration of a reactant or product or (3) by a change in the volume of the container. The outcomes of these kinds of changes can be predicted by **LeChatelier's principle** which states: A change in any of the factors that determine the equilibrium conditions of a system causes the system to change in a manner that will counteract the effect of the change.

Effect of Temperature Change on Equilibria

As the temperature of a system at equilibrium is raised by adding heat energy, the system reacts in the direction that absorbs heat to reestablish a state of equilibrium. This means an increase in temperature always causes the system to shift in the endothermic direction to achieve the new state of equilibrium, because heat can be thought of as a reactant in an endothermic change. Conversely, lowering the temperature of a system at equilibrium causes a shift in the heat releasing or exothermic direction. ✶ See the margin summary on p. 282 in the text, and try preparing a "flash card" stating the information of that summary and summarizing the effect of temperature changes on systems at equilibrium and memorizing it.

A change in temperature is the only kind of change that causes a change in the value of K. If the reaction is endothermic in the forward direction, it is enhanced in that direction by an increase in temperature. This means the concentrations of the products of an endothermic reaction increase and those of the reactants decrease with an increase in temperature, and *the value of K,* which is a reflection of the ratio of product concentrations over reactant concentrations, *increases with increasing temperature for an endothermic reaction. Conversely, the value of K for an exothermic reaction decreases with increasing temperature,* because the increase in temperature opposes the tendency of the exothermic reaction to give off energy as heat. ✱ Try preparing a "flash card" which summarizes the effect of an increase in temperature on an endothermic reaction and on an exothermic reaction and also gives a brief explanation for these effects and memorize it.

Effect of the Addition or Removal of a Reactant or a Product

When the concentration of a reactant or product is changed from its equilibrium value at a given temperature, the reaction shifts to a new equilibrium position for which the reaction quotient is still equal to K. In summary,

- addition of a reactant or removal of a product causes Q to temporarily be less than K and the reaction to proceed in the product direction to reestablish equilibrium.

- removal of a reactant or addition of product causes Q to temporarily be greater than K and the reaction to proceed in the reactant direction to reestablish equilibrium.

Ex. 16.9 in the text illustrates the procedure that is used to calculate the equilibrium concentrations that are obtained following a change in concentration.

The Effect of Volume Change on Gas Phase Equilibria

When the volume of the container containing a gaseous system at equilibrium is decreased, there is a pressure increase that is counteracted by a shift to the reaction side that contains the fewer number of molecules of gases. Conversely, an increase in the volume of the container causes a pressure decrease that is counteracted by a shift to the reaction side that contains the greater number of molecules of gases.

The commercial impact of changes affecting chemical equilibrium is illustrated in the "Current Issues in Chemistry" that describes the production of ammonia, which is used extensively as a fertilizer. The effects of disturbances on Equilibrium and K are also summarized in Table 16.2 in the text. ✱ Try preparing a "flash card" summarizing the effects of changes in concentration on equilibrium and a second "flash card" summarizing the effects of changes in volume on equilibrium and memorizing them.

16.7 Equilibrium, Kinetics and Reaction Mechanisms

Reaction Rates and Equilibrium

When the rates of the forward and reverse reactions are equal, there is no net change in concentrations, as illustrated in Fig.16.9 in the text. **The forward and reverse reactions continue to occur, but do so at equal rates, and the system is at equilibrium.**

Reaction Mechanisms and Equilibrium

All the reaction mechanisms discussed in Section 15.6 involved a rate determining elementary step as the first step followed by one or more elementary fast steps. However, **many reactions are believed to use a fast and reversible elementary step as a first step followed by a rate determining elementary step and possibly some fast steps.** For example, the reaction

$$2 \; NO(g) \; + \; O_2(g) \; \rightarrow \; 2 \; NO_2(g) \tag{16k}$$

obeys the rate equation

$$Rate \; = \; k[NO]^2[O_2] \tag{16\ell}$$

and **is thought to follow a two-step mechanism, because a one-step termolecular gas phase reaction is not likely to occur. It is thought to follow:**

Elementary step 1: Fast and reversible equilibrium

$$NO(g) \; + \; O_2(g) \; \underset{k_{-1}}{\overset{k_1}{\rightleftharpoons}} \; OONO(g) \tag{16m}$$

Elementary step 2: Slow and rate determining

$$NO(g) \; + \; OONO(g) \; \underset{k_2}{\rightarrow} \; 2 \; NO_2(g) \tag{16n}$$

Overall: $2 \; NO(g) \; + \; O_2(g) \; \rightarrow \; 2 \; NO_2(g)$ (16o)

The rate equation predicted by the rate determining step is:

$$\text{Rate} = k_2[\text{NO}][\text{OONO}] \tag{16p}$$

but this contains the concentration of the intermediate OONO whose concentration cannot be measured. We can therefore use the fast and reversible first step to obtain an expression for [OONO] that can be substituted into the predicted rate equation.

Because the second step is slow compared to the reverse of the first step, much of the OONO produced by the first step reverts to NO and O_2 before it can be consumed in the second step. This enables the first step to reach a state of equilibrium before any significant amount of OONO is consumed in the second step. This state of equilibrium remains throughout the lifetime of the overall reaction and is only broken when the overall reaction is nearly complete.

Because the state of chemical equilibrium involves opposing reactions occurring at equal rates, we can write

Rate of forward reaction = Rate of reverse reaction

$$k_1[\text{NO}][O_2] = k_{-1}[\text{OONO}] \tag{16q}$$

solve for [OONO] and substitute into the predicted rate equation. This gives:

$$[\text{OONO}] = \frac{k_1}{k_{-1}}[\text{NO}][O_2] \tag{16r}$$

and

$$\text{Rate} = k_2[\text{NO}]\left(\frac{k_1}{k_{-1}}[\text{NO}][O_2]\right) \tag{16s}$$

$$= \frac{k_2 k_1}{k_{-1}}[\text{NO}]^2[O_2] \tag{16t}$$

Because the first term involves just constants and is always equal to a constant at a given temperature, we can write

$$\text{Rate} = k'[\text{NO}]^2[O_2] \tag{16u}$$

which agrees with the experimentally observed rate equation and the overall stoichiometry for the reaction. ***Thus, we have postulated a reasonable mechanism. However, it is not the only mechanism that is consistent with the experimental rate law,*** as seen in Ex. 16.10 in the text. In that mechanism, the fast and reversible first step involves the dimerization of NO

$$NO(g) + NO(g) \rightleftharpoons N_2O_2(g) \qquad (16v)$$

which is plausible because each NO molecule has a single unpaired electron on the N atom. **However, there is experimental evidence for the formation of OONO during the course of the reaction, and the mechanism involving formation of OONO as an intermediate is favored.**

An Alternative Method for Solving for the Concentration of An Intermediate Formed in a Rapid and Reversible Equilibrium Step:

An expression for [OONO] can also be obtained by writing the equilibrium constant expression for elementary step 1 and solving for [OONO]. This gives

$$K_c = \frac{[OONO]}{[NO][O_2]} \qquad (16w)$$

and

$$[OONO] = K_c[NO][O_2] \qquad (16x)$$

so,

$$Rate = k_2[NO](K_2[NO][O_2]) \qquad (16y)$$

$$= k_2K_c[NO]^2[O_2] \qquad (16z)$$

$$= k'[NO]^2[O_2] \qquad (16aa)$$

Comparing Expressions 16r and 16x , we see that

$$K_c = \frac{k_1}{k_{-1}} \qquad (16ab)$$

as derived in the text, which uses a combination of the approaches used here for working with fast and reversible first-step mechanisms. **This means an equilibrium constant for a reaction is equal to the ratio of the rate constants for the forward and reverse reactions and helps us understand why the ratio of product concentrations over reactant concentrations, each raised to the power of their stoichiometric coefficient, is a constant at any given temperature.**

16.8 Key Expressions

General equation equilibrium constant expression:

For the general reaction

$$aA + bB \rightleftharpoons cC + dD$$

the equilibrium constant expression is

$$K_c = \frac{[C]^c[D]^d}{[A]^a[B]^b}$$

where the molar concentrations pertain to chemical equilibrium.

General equation reaction quotient expression:

For the general reaction

$$aA + bB \rightleftharpoons cC + dD$$

the reaction quotient expression is

$$Q_c = \frac{[C]^c[D]^d}{[A]^a[B]^b}$$

where the molar concentrations pertain to any point in the chemical reaction. At equilibrium, $Q_c = K_c$.

Test for validity of approximation expression:

$$\frac{x}{\text{smallest initial concentration}} \cdot 100\% \text{ must be } \leq 5\%$$

16.9 Hints for Answering Questions and Solving Problems

Writing equilibrium constant and reaction quotient expressions:

Place the molar concentrations of products over the molar concentrations of reactants, and raise the concentration of each to the power of its stoichiometric coefficient. Omit terms for solids and pure liquids used as solvents, including water for reactions conducted in aqueous solutions.

Constructing ICE tables:

Express the changes in terms of a single unknown, x, and try to avoid using fractions. For example, for the reaction $2A \rightleftharpoons B + 2C$, let x be the amount of B formed and $-2x$ and $+2x$ be the respective changes for A and C.

Conducting calculations for equilibrium concentrations:

Check to see if the perfect-square method can be used. If not, try the method of neglecting x with respect to nonzero initial concentrations, and check to see if the approximation is valid.

CHAPTER 17

PRINCIPLES OF REACTIVITY:
THE CHEMISTRY of ACIDS AND BASES

Acids and bases are everywhere, and acid-base reactions are one of the most important types of reactions. In this chapter, we will consider the Arrhenius, Bronsted and Lewis theories of acids and bases but concentrate most on the Bronsted theory. We will consider the pH scale for expressing acidity and consider the relative strengths of acids and bases and the relation between chemical bonding, molecular structure and acid strength. We will also consider the equilibria of weak acids and bases and the acid-base nature of salt solutions in considerable detail. However, the area of acid-base chemistry is an area of general chemistry in which students commonly have difficulty, so you should spend some extra time mastering these important concepts.

17.1 Acids, Bases and Arrhenius

Acids, (Alkalies) Bases and Salts

- The word "**acid**" comes from the Latin word *acidus*, meaning sour. *Acids are characterized by their sour taste, their ability to be corrosive, redden blue vegetable colors and lose all of these properties when reacted with alkalies* (4.4).

- The term "**alkali**" is derived from the Arabic word for the ashes that come from burning certain plants. Potash (potassium carbonate) is one of the products formed by this process, and because water solutions of potash feel soapy and taste bitter, the term "alkali" was later applied to other substances having those properties. It was also recognized that acids react with "alkalies" to form salts, so "alkalies" were considered to be the "bases" for these salts, and the term "**base**" became used to describe "alkalies." *Bases are characterized by their soapy/slippery solutions, ability to restore vegetable colors reddened by acids and ability to react with acids to form salts (4.4).*

- The term "**salt**" is used to describe any ionic compound whose cation could come from a base and whose anion could come from an acid in an acid-base reaction (4.8).

---

* Try to prepare a "flash card" listing the general properties of acids and bases and memorizing them.

Arrhenius's Definition of Acids and Bases

According to the acid-base model put forth by the Swedish chemist Svante Arrhenius (1859-1927),

- an **acid** is a substance that contains hydrogen and produces hydrogen ions (H^+) by ionic dissociation in water.

$$HB(aq) \rightarrow H^+(aq)\ B^-(aq) \tag{17a}$$

- a **base** is a substance that produces hydroxide (OH^-) ions in water.

$$MOH(aq) \rightarrow M^+(aq) + OH^-(aq) \tag{17b}$$

$$M(OH)_2(aq) \rightarrow M^{2+}(aq) + 2\ OH^-(aq) \tag{17c}$$

$$NH_3(aq) + H_2O(\ell) \rightarrow NH_4^+(aq) + OH^-(aq) \tag{17d}$$

- an acid and a base react to produce a **salt** and water.

$$HCl(aq) + NaOH(aq) \rightarrow NaCl(aq) + H_2O(\ell) \tag{17e}$$

$$2\ HNO_3(aq) + Fe(OH)_2(aq) \rightarrow Fe(NO_3)_2(aq) + 2\ H_2O(\ell) \tag{17f}$$

* Try preparing a "flash card" giving the Arrhenius definitions of an acid and a base and the general form of an Arrhenius acid-base reaction, namely acid + base → salt + water, and memorizing it.

Arrhenius's concept is limited to aqueous solutions, because it refers to ions (H^+ and OH^-) derived from water. More general concepts of acids and bases are discussed later in the chapter.

17.2 The Hydronium Ion and Water Autoionization

A H^+ ion in water is usually represented as the **hydronium ion**, H_3O^+, as described in the "Closer Look" on p. 173 in the text. *Two water molecules can react with each other to produce a hydronium ion and a hydroxide ion by proton transfer from one water molecule to the other* in a so-called **autoionization** reaction.

$$H_2O(\ell) + H_2O(\ell) \rightleftharpoons H_3O^+(aq) + OH^-(aq) \qquad (17g)$$

This reaction, demonstrated by Friedrich Kohlrausch (1840-1910), **occurs in pure water and in aqueous solutions** and is therefore the cornerstone of acid-base behavior in aqueous solutions.

17.3 The Bronsted Concept of Acids and Bases

In 1923, Johannes Bronsted of Copenhagen, Denmark and Thomas Lowry of Cambridge, England independently suggested a new concept of acid and base behavior. **They proposed that:**

 • an **acid** is any substance that can donate a proton to another substance, so **acids can be neutral molecules, cations or anions.**

$$HNO_3(aq) + H_2O(\ell) \rightarrow H_3O^+(aq) + NO_3^-(aq) \qquad (17h)$$

$$NH_4^+(aq) + H_2O(\ell) \rightarrow H_3O^+(aq) + NH_3(aq) \qquad (17i)$$

$$H_2PO_4^-(aq) + H_2O(\ell) \rightarrow H_3O^+(aq) + HPO_4^-(aq) \qquad (17j)$$

 • a **base** is any substance that can accept a proton from another substance, so **bases can be neutral molecules or anions.**

$$NH_3(aq) + H_2O(\ell) \rightleftharpoons NH_4^+(aq) + OH^-(aq) \qquad (17k)$$

$$PO_4^{3-}(aq) + H_2O(\ell) \rightleftharpoons HPO_4^{2-}(aq) + OH^-(aq) \qquad (17\ell)$$

✶ Try preparing a "flash card" giving the Bronsted definitions of an acid and a base and memorizing it.

Acids capable of donating just one proton are called **monoprotic acids** whereas acids capable of donating two or more protons are called **polyprotic acids**. You are familiar with some examples of these types of acids from your study of Chapter 4 (Table 4.1), and additional examples of polyprotic acids are given in Table 17.1 in the text. ✶ Try preparing a "flash card" listing the common examples of monoprotic, diprotic and triprotic acids given in Tables 4.1 and 17.1 in the text and memorizing the names and formulas of these acids by category.

Anions of polyprotic acids having charges of 2- or higher can accept more than one proton and thus act as **polyprotic bases**. Common examples include CO_3^{2-}, $C_2O_4^{2-}$, S^{2-}, SO_4^{2-}, HPO_4^{2-} and PO_4^{3-}.

Molecules and ions that can behave as either a Bronsted acid or base are called **amphiprotic** substances. **Water is the most common amphiprotic substance.** Water acts as a base in reactions with acids, such as the reaction with HNO_3 in Expression 17h, as an acid in its reactions with bases, such as the reaction with NH_3 in Expression 17k, and as both an acid and a base in its autoionization reaction given in Expression 17g. **Other important amphiprotic substances include the intermediate forms of polyprotic acids that have not yet had all of the ionizable hydrogen atoms removed,** such as HS^-, $H_2PO_4^-$ and HPO_4^-.

Conjugate Acid-Base Pairs

Substances that differ by the presence of one H^+ unit in their chemical formulas are called a **conjugate acid-base pair**. We say that CO_3^{2-} is the conjugate base of HCO_3^-, and that HCO_3^- is the conjugate acid of CO_3^{2-}. However, HCO_3- is amphiprotic, so HCO_3^- is also the conjugate base of H_2CO_3, and H_2CO_3 is the conjugate acid of HCO_3^-.

Every acid-base reaction involving H^+ transfer involves two conjugate acid-base pairs, as shown in Table 17.2 in the text. **The general form of a Bronsted acid-base reaction is:**

$$Acid\ 1 + Base\ 2 \rightleftharpoons Acid\ 2 + Base\ 1 \qquad (17m)$$

where Acid 1 and Base 1 are a conjugate acid-base pair and Base 2 and Acid 2 are a conjugate acid-base pair. For example, in the reaction in Expression 17h, HNO_3 and NO_3^- are a conjugate acid-base pair, and H_2O and H_3O^+ are a conjugate acid-base pair. Similarly, in the reaction in Expression 17k, NH_3 and NH_4^+ are a conjugate acid-base pair, and H_2O and OH^- are a conjugate acid-base pair. ✶ Try preparing a "flash card" containing the general form of a Bronsted acid-base reaction and identifying the members of the conjugate pairs involved and memorizing it.

Relative Strengths of Acids and Bases

Some acids are better proton donors in water than other acids (4.4), and some bases are better proton acceptors than others. For example, **a dilute solution of hydrochloric acid is nearly 100% ionized, and hydrochloric acid is considered to be a strong Bronsted acid**. This means that a 0.1 M aqueous solution of HCl actually consists of 0.1 M H_3O^+ and 0.1 M Cl^- and few, if any, molecules of HCl. On the other hand, the ionization of acetic acid is so slight that a 0.1 M aqueous solution of acetic acid consists of about 0.001 M H_3O^+, 0.001 M $CH_3CO_2^-$ and 0.099 M CH_3CO_2H. That is, **only about 1% of CH_3CO_2H molecules ionize in a 0.1 M CH_3CO_2H solution, and acetic acid is considered to be a weak acid.** These concepts were illustrated with the use of chemical equations in 4.4, and you should review them as needed.

The oxide ion is too strong a base to exist in aqueous solution. It reacts completely with water to produce hydroxide ions

$$O^{2-}(aq) + H_2O(\ell) \rightarrow 2\ OH^-(aq) \tag{17n}$$

so the dissolving of lithium oxide is written as

$$Li_2O(s) + H_2O(\ell) \rightarrow 2\ Li^+(aq) + 2\ OH^-(aq) \tag{17o}$$

On the other hand, aqueous ammonia and the carbonate ion only produce a very small concentration of hydroxide ions and are classified as weak Bronsted bases.

$$NH_3(aq) + H_2O(\ell) \rightleftharpoons NH_4^+(aq) + OH^-(aq) \tag{17p}$$

$$CO_3^{2-}(aq) + H_2O(\ell) \rightleftharpoons HCO_3^-(aq) + OH^-(aq) \tag{17q}$$

Strong acids necessarily have very weak conjugate bases. For example, aqueous HCl is a strong acid because it has a strong tendency to donate a proton to water and form its conjugate base Cl^-, so Cl^- must have little tendency to retain the proton. *In general, the stronger the acid, the weaker its conjugate base, so*

- very weak acids have strong conjugate bases,

- weak acids have weak conjugate bases and

- strong acids have very weak conjugate bases.

and the converse is, the stronger the base, the weaker its conjugate acid. Thus, the differing extent of the reactions of HCl and CH_3CO_2H with water illustrate an important principle in Bronsted acid-base theory: *all proton transfer reactions proceed predominantly from the stronger acid-base pair to the weaker acid-base pair.* ✷ Try preparing two separate "flash cards," one listing the three general categories of relative strengths for conjugate acid-base pairs and the other stating the predominant direction of Bronsted acid-base reactions, and memorizing them.

Acids and their conjugate bases can be ordered on the basis of their relative tendencies to donate and accept protons, as in Table 17.3 in the text. These entries can then be used to predict whether the equilibrium lies predominantly to the left or the right in an acid-base reaction.

Using Relative Acid-Base Strengths to Predict the Predominant Direction of Acid-Base Reactions

In general, when acids are arranged in an order of decreasing acid strength, as in Table 17.3, *the stronger acids located in the top left of the table will react predominantly with the stronger bases located in the bottom right of the table.* Thus, to predict the predominant direction of the reaction between CH_3CO_2H and NaCN, we first note that NaCN is a soluble salt that exists as Na^+ and CN^- in aqueous solution and that the reaction involves the transfer of a proton from CH_3CO_2H to CN^-

$$CH_3CO_2H(aq) + CN^-(aq) \rightleftharpoons CH_3CO_2^-(aq) + HCN(aq) \qquad (17r)$$

Secondly, we note that CH_3CO_2H is above HCN on the left in Table 17.3 and is therefore a stronger acid than HCN and that CN^- is below $CH_3CO_2^-$ on the right in Table 17.3 and is therefore a stronger base than $CH_3CO_2^-$, so we predict the predominant direction of reaction is in the forward direction. In other words, we predict the reaction is product-favored.

17.4 Strong Acids and Bases

The hydronium ion is actually the strongest acid that can exist in water. All acids listed above H_3O^+ on the top left side of Table 17.3 react completely with water to form H_3O^+ and their respective conjugate bases and thus appear to be the same strength in water. We refer to this as the **levelling effect** of water, because the acid strength of these acids is brought to the level of H_3O^+. However, you should note that this only applies to removal of the first proton of H_2SO_4, since HSO_4^- is a negatively charged species that is a weaker acid than H_3O^+ (Table 17.4).

Similarly, the hydroxide ion is actually the strongest base that can exist in water. All bases below OH^- on the bottom right side of Table 17.3 react completely with water to form OH^- and their respective conjugate acids and thus appear to be the same strength in water. They are brought to the level of OH^- in water. For example, the hydride, H^-, ion is too strong a base to exist in water, and the vigorous reaction of H^- with H_2O

$$H^-(aq) + H_2O(\ell) \rightarrow H_2(g) + OH^-(aq) \qquad (17s)$$

is illustrated in Fig. 17.2 in the text. However, *the most common examples of strong bases are the soluble metal oxides and hydroxides of elements of Groups IA and IIA* (4.4). ✳ Try preparing a "flash card" summarizing what is meant by the levelling effect of the solvent as it applies to water and include an example of a chemical equation pertaining to the levelling of a strong acid and an example of a chemical equation pertaining to the levelling of a strong base and memorizing it.

17.5 Weak Acids and Bases

The majority of acids and bases are weak. The relative strength of an acid or base can be expressed quantitatively with an equilibrium constant.

For the general weak acid HA, we can write

$$HA(aq) + H_2O(\ell) \rightleftharpoons H_3O^+(aq) + A^-(aq) \qquad (17t)$$

and

$$K_a = \frac{[H_3O^+][A^-]}{[HA]} \qquad (17u, 17.1)$$

where K has a subscript "a" to indicate that it is an equilibrium constant for a weak acid in water, and the concentration of water is omitted from the equilibrium expression in keeping with the principles developed in Section 16.2. The value of K is less than 1, indicating the product of the equilibrium concentration of the hydronium ion and the equilibrium concentration of the conjugate base of the acid is less than the equilibrium concentration of the acid. In other words, the reaction is reactant-favored.

For the general weak base B, we can write

$$B(aq) + H_2O(\ell) \rightleftharpoons BH^+(aq) + OH^-(aq) \qquad (17v)$$

and

$$K_b = \frac{[BH^+][OH^-]}{[B]} \qquad (17w, 17.2)$$

The value of K_b is also less than 1, because the reaction is reactant-favored. ∗ Try preparing a "flash card" summarizing the general equations for the reactions of weak acids and weak bases with water and the resulting K_a and K_b expressions and also include a specific example of a reaction of a weak acid and a specific example of a reaction of a weak base and memorize it.

The relative strengths of acids and their respective conjugate bases are given in terms of their K_a and K_b values in Table 17.4, which is organized like Table 17.3. Thus, *the strongest acids having the highest K_a values are listed at top of the left side, and the strongest bases having the highest K_b values are listed* at the bottom on the right side. Once again, notice that the **stronger the acid, the weaker the conjugate base. However, we are now in position to classify acids and bases on the basis of their K_a and K_b values as follows:**

Acid Strength	K_a	Conjugate Base Strength	K_b
Strong	>1	very weak	$<10^{-16}$
Weak	1 to 10^{-16}	weak	1 to 10 $^{-16}$
Very weak	$<10^{-16}$	strong	>1

✴ Try preparing a "flash card" summarizing this information and memorizing it.

Some Weak Acids

Neutral Molecules as Weak Acids:

The vast majority of neutral molecules with an ionizable hydrogen atom are weak Bronsted acids. See Fig. 17.3 and Table 17.4 in the text for examples.

Cations as Weak Acids:

The ammonium ion, NH_4^+, is the conjugate acid of the weak base ammonia, NH_3, and therefore acts as a weak acid.

$$NH_4^+(aq) + H_2O(\ell) \rightleftharpoons H_3O^+(aq) + NH_3(aq) \qquad (17x)$$

In addition, ***hydrated metal cations of high charge and small size act as weak acids.*** For example, $Al(H_2O)_6^{3+}$ is almost as strong an acid as CH_3CO_2H (Table 17.4) and is a threat to the environment, because acid rain can leach Al^{3+} ions from the soil.

$$Al(H_2O)_6^{3+}(aq) + H_2O(\ell) \rightleftharpoons H_3O^+(aq) + Al(H_2O)_5(OH)^{2+}(aq) \qquad (17y)$$

Anions as Weak Acids:

Six anions containing ionizable H atoms are listed as Bronsted acids in Table 17.4. The strongest of these acids is the hydrogen sulfate ion:

$$HSO_4^-(aq) + H_2O(\ell) \rightleftharpoons H_3O^+(aq) + SO_4^{2-}(aq) \qquad (17z)$$

Some Weak Bases

Neutral Molecules as Weak Bases:

Ammonia is probably the most common example of a neutral

molecule that acts as a weak base, as shown in Expression 17p above. **Ammonia is also the parent compound of a large series basic compounds called amines** (11.7). **Amines** are formed by substituting various alkyl groups for the hydrogen atoms of ammonia, such as the methyl groups in methylamine, CH_3NH_2; dimethylamine, $(CH_3)_2NH$; and trimethylamine, $(CH_3)_3N$. **The key to each of these compounds acting as weak bases is the lone pair of electrons on the N atom that can accept a proton by forming a coordinate covalent bond with the H^+.**

Anions as Weak Bases

We have already seen PO_4^{3-} and CO_3^{2-} acting as weak bases in water in Expressions 17ℓ and 17q.

17.6 Water and the pH Scale

The stronger a Brønsted acid or base, the larger the concentration of H_3O^+ or OH^- for a given concentration of acid or base, respectively. The pH scale gives us a way to measure these concentrations quantitatively and enables us to compare acid and base strengths.

The Water Ionization Constant, K_w

The autoionization of water produces hydronium ions and hydroxide ions. However, the equilibrium position for the auto-ionization reaction (17g)

$$H_2O(\ell) + H_2O(\ell) \rightleftharpoons H_3O^+(aq) + OH^-(aq)$$

lies far to the left, because H_3O^+ is a much stronger acid than H_2O and OH^- is a much stronger base than H_2O (Table 17.4 in the text). Indeed, only 2 out of approximately 10^9 (a billion) water molecules are ionized at any given time at 25°C. This can be expressed more quantitatively in terms of the equilibrium expression

$$K = \frac{[H_3O^+][OH^-]}{[H_2O]^2} \tag{17ab}$$

However, the concentration of H_2O is always 55.5 M, no matter how much water is present, and the concentration of H_2O can be incorporated into the equilibrium constant (16.2) giving

$$K[H_2O]^2 = [H_3O^+][OH^-]$$

or simply

$$K_w = [H_3O^+][OH^-] \qquad\qquad (17ac, 17.3)$$

which is applicable to both pure water and aqueous solutions.
The constant K_w is known as the **ionization constant of water.**

In pure water, H_3O^+ and OH^- are formed in equal amounts by the autoionization of water. Careful electrical conductivity measurements of pure water show that $[H_3O^+] = [OH^-] = 1.0 \times 10^{-7}$ M at $25°C$, so

$$K_w = [H_3O^+][OH^-] = (1.0 \times 10^{-7})(1.0 \times 10^{-7}) = 1.0 \times 10^{-14} \quad (17ad)$$

at $25°C$. Because the hydronium and hydroxide concentrations in pure water are both 1.0×10^{-7} M at $25°C$, the water is said to **neutral.** However, K_w **is temperature dependent,** as indicated by the table in the margin of p. 811 in the text, **and the concentrations of H_3O^+ and OH^- found in pure water and neutral solutions at other temperatures vary with temperature.** They can be calculated using $x^2 = K_w$, where $x = [H_3O^+] = [OH^-]$, with K_w at any given temperature. ✳ Try preparing a "flash card" showing the water ionization constant expression (17ac) and its value of 1.0×10^{-14} at $25°C$ and memorizing it.

Adding an acid to water raises the $[H_3O^+]$. To oppose this addition, LeChatelier's principle (16.7) predicts H_3O^+ will react with OH^- to form water and lower the $[OH^-]$ until the product $[H_3O^+][OH^-]$ again equals 1.0×10^{-14} at $25°C$. Similarly, adding a base to water raises the $[OH^-]$, and LeChatelier's principle predicts OH^- will react with H_3O^+ to form water and lower the $[H_3O^+]$ until the product $[H_3O^+][OH^-]$ again equals 1.0×10^{-14} at $25°C$. Thus, for aqueous solutions at $25°C$, we can say that

- In a **neutral solution**, $[H_3O^+] = [OH^-]$. $\qquad$ (17ae)

  Both are equal to 1.0×10^{-7} M.

- In an **acidic solution**, $[H_3O^+] > [OH^-]$. $\qquad$ (17af)

  $[H_3O^+] > 1.0 \times 10^{-7}$ M and $[OH^-] < 1.0 \times 10^{-7}$.

- In a **basic solution**, $[H_3O^+] < [OH^-]$. $\qquad$ (17ag)

  $[H_3O^+] < 1.0 \times 10^{-7}$ M and $[OH^-] > 1.0 \times 10^{-7}$.

✳ Try preparing a "flash card" summarizing these characteristics of neutral, acidic and basic solutions and memorizing it.

Ex. 17.2 in the text shows the concentration of H_3O^+ or OH^- from the autoionization of water can usually be neglected when calculating the $[H_3O^+]$ or $[OH^-]$ for solutions of strong acids or bases of moderate concentration (i.e. $\geq 10^{-5}$ M).

The Connection Between the Ionization Constants for Water, a Weak Acid and Its Conjugate Base

The entries in Table 17.4 in the text clearly indicate **there is an inverse relation between K_a and K_b for members of conjugate pairs; as K_a decreases, K_b increases and vice versa.** Now that the concept of K_w has been introduced, **this relation can be expressed mathematically as**

$$K_a \cdot K_b = K_w \qquad\qquad (17ah, 17.4)$$

The basis for this expression can be seen by considering the relation between the acid ionization expression for HCN and the base ionization expression for CN^- given on p. 813 in the text.

✶ Try preparing a "flash card" showing Expression 17ah and memorizing it.

The pH Scale

The **pH scale** is a widely used logarithmic scale that is used to express hydronium ion concentrations and acidity and avoid the use of very small numbers and exponential notation. The **pH** of a solution is defined as the negative of the base-10 logarithm (log) of the hydronium ion concentration.

$$pH = -\log[H_3O^+] \qquad\qquad (17ai, 17.5)$$

The higher the $[H_3O^+]$, the lower the pH and the more acidic the solution. Similarly, the higher the $[OH^-]$, the lower the pOH and the more basic the solution, where **pOH** is defined as the negative of the base-10 logarithm of the hydroxide ion concentration.

$$pOH = -\log[OH^-] \qquad\qquad (17aj, 17.6)$$

In pure water at 25°C $[H_3O^+] = [OH^-] = 1.0 \times 10^{-7}$, so pH = 7.00 and pOH = 7.00. Indeed

$$pH + pOH = 14 \qquad\qquad (17ak, 17.7)$$

for all aqueous solutions at 25°C, because this relation is based on the value of K_w, as shown on p. 814 in the text. **At other temperatures,**

$$pH + pOH = -\log K_w \qquad\qquad (17a\ell)$$

but you can assume the temperature is 25°C unless told otherwise.

Note: It is common to express pH and pOH values to two significant figures. This corresponds to two decimal places, because numbers to the left of the decimal in logarithmic quantities are related to the power of ten that is used to express the number in scientific notation. In addition, it is common to take the sum of pH and pOH at 25°C to be exactly 14, because the origin of this number is the exact exponent of the power of ten that appears in the value of K_w at 25°C.

Exs. 17.3 and 17.4 in the text illustrate working with these concepts, and "Problem Solving Tips and Ideas" 17.2 also addresses "Calculating and Using pH." *Notice that given pH or pOH, it is possible to calculate [H₃O⁺] or [OH⁻] using*

$$[H_3O^+] = \text{antilog}(-pH) = 10^{-pH} \qquad (17am)$$

or

$$[OH^-] = \text{antilog}(-pOH) = 10^{-pOH} \qquad (17an)$$

✷ Try preparing a "flash card" summarizing Expressions 17 ai, aj, am, an and ak and memorizing it.

The relation between acidity, basicity, pH and pOH is illustrated graphically on p. 816 in the text. ✷ Try preparing a "flash card" summarizing the information of this chart and memorizing it.

Determining pH

The pH of a solution can be determined in an approximate manner using an **acid-base indicator**, a substance which changes color in some known pH range. *These substances can exist in conjugate acid and base forms having different colors*. Common litmus paper, for example, is impregnated with a natural plant juice that is red in solutions more acidic than about pH = 5 but blue when the pH exceeds about 8.2 in basic solution. The pH ranges of other indicators are given in Fig. 17.7 in the text.

A modern pH meter can be used to determine the pH of a solution more accurately. A pH meter is shown in Fig. 17.8 in the text.

17.7 Equilibria Involving Weak Acids and Bases

Calculating K_a or K_b from Initial Concentrations and Measured pH

The K_a and K_b values given in Table 17.4 and in Appendices F and G were all determined by experiment. **Ex. 17.5 in the text illustrates the procedure for using the pH of an acid solution of given concentration to calculate K_a. Notice that this involves constructing an ICE table, calculating $[H_3O^+]$ from pH using Expression 17am, calculating the equilibrium concentrations of the acid, hydronium ion and conjugate base of the acid and substituting into the K_a expression to calculate K_a.** ∗ Try preparing a "flash card" summarizing this procedure and memorizing it. Also try practicing this type of calculation until you are comfortable doing them and can do them quickly. **In addition, realize that a similar procedure can be used to calculate K_b, given pH or pOH and initial weak base concentration.** ∗ Try preparing a "flash card" summarizing the procedure you would use to calculate K_b given the pH or pOH of a solution of a weak base of a given molarity and memorizing it. Also look for examples of this in the end-of-chapter "Study Questions" and try practicing this type of calculation until you are comfortable doing them and can do them quickly.

There is another important item to note concerning the solution to Ex. 17.5. That is, **in this case the amount of acid ionized is so small that it is negligible compared to initial concentration of the acid, and the equilibrium concentration of acid is essentially equal to the initial concentration of acid. Thus, Expression 17t**

$$HA(aq) + H_2O(\ell) \rightleftharpoons H_3O^+(aq) + A^-(aq)$$

leads to

$$K_a = \frac{[H_3O^+][A^-]}{[HA]_o - [H_3O^+]} \approx \frac{[H_3O^+][A^-]}{[HA]_o} \qquad (17ao)$$

where $[HA]_o$ is the initial concentration of the weak acid. **Neglecting $[H_3O^+]$ with respect to $[HA]_o$ is analogous to the** neglecting of x that was done in 16.5, and error analysis shows that this **can be done whenever $[HA]_o \geq 100 \cdot K_a$.** That is, $[HA]_{equil}$ is effectively equal to $[HA]_o$, whenever $[HA]_o \geq 100 \cdot K_a$.

Calculating Equilibrium Concentrations and pH from Initial Concentrations and K_a or K_b

The pH of a solution of a weak acid or weak base of known concentration can be calculated using the K_a or K_b of the acid or base, respectively. The procedure for doing so is illustrated in Exs. 17.6-8 in the text. **Notice that the procedure for calculating the equilibrium concentration of H_3O^+ or OH^- is similar to that used for solving Ex. 16.8; that is,**

• constructing an appropriate ICE table,

• setting-up the equilibrium constant expression in terms of the equilibrium concentrations of the ICE table,

• solving for x by neglecting x with respect to the initial concentration of acid or base, if $[HA]_o \geq 100 \cdot K_a$ or $[B]_o \geq 100 \cdot K_b$, or if it is not, solving for x using the quadratic method (16.5) and

• then going on to calculate pH using $[H_3O^+] = x$ **or** pOH using $[OH^-] = x$ and pH using pH = 14 - pOH.

The percentage of weak acid or base ionized can furthermore be calculated using

$$\%ionized = \frac{quantity\ ionized}{initial\ concentration} \cdot 100\% \qquad (17ap)$$

$$= \frac{x}{[HA]_o} \cdot 100\% \ \underline{or} \ \frac{x}{[B]_o} \cdot 100\%$$

✶ Try preparing separate "flash cards" summarizing the procedure for calculating the hydronium ion concentration and pH for a solution of a weak acid, the procedure for calculating the hydroxide ion concentration and pH for a solution of a weak base and the procedure for calculating percent ionization and memorizing them. Also try doing these kinds of calculations until you are comfortable doing them and can do them quickly.

17.8   Acid-Base Properties of Salts: Hydrolysis

A **salt**, you recall, is an ionic compound whose cation could come from a base and whose anion could come from an acid in an acid-base reaction(4.8). ***Many salts readily dissolve in water to give acidic or basic aqueous solutions by hydrolysis.*** A **hydrolysis reaction** is said to have occurred whenever a salt dissolves in water and produces changes in the concentrations of the H_3O^+ and OH^- ions of the water. ***Ions can be placed into four categories in terms of their reactions with water.***

• Cations which are conjugate acids of weak bases and hydrated metal cations of high charge and small size react with water to produce acidic solutions. See, for example, Expressions 17x and 17y.

• Anions which are conjugate bases of weak acids react with

water to produce basic solutions. See, for example, Expressions 17ℓ and 17q.

• Anions which are conjugate bases of weak polyprotic acids and contain ionizable hydrogen atoms are amphiprotic and can act as either weak acids or weak bases to produce acidic or basic solutions, depending on their K_a and K_b values. See, for example, the discussion of HCO_3^- on p. 829 in the text.

• Cations which come from strong bases and anions which come from strong acids do not react appreciably with water and thus do not affect the acidity or basicity of solutions.

✶ Try preparing four separate "flash cards" summarizing the information for each of these categories and memorizing them.

The categories of ions given above can be used to predict whether solutions of given salts will be acidic, neutral or basic, as is done on p. 829 in the text. *In some cases, such as $NH_4CH_3CO_2$ and NH_4F, the cation acts as a weak acid and the anion acts as a weak base. In these cases, the relative values of the K_a of the cation and the K_b of the anion determine whether the solution will be acidic, neutral or basic. If,*

K_a cation > K_b anion, the solution will be acidic.
K_a cation = K_b anion, the solution will be neutral.
K_a cation < K_b anion, the solution will be basic.

This leads us to the following summary:

Cation	Anion	pH of the Solution
From strong base	From strong acid	=7 (neutral)
From strong base	From weak acid	>7 (acidic)
From weak base	From strong acid	<7 (basic)
From weak base	From weak acid	Depends on K_a and K_b.

✶ Try preparing a "flash card" giving this summary and the three possible combinations for K_a cation and K_b anion given just above it and memorizing it. However, you should also be prepared to give explanations for these observations in terms of the expected hydrolysis reactions and be able to write equations to support your explanations. If you have any doubt about whether you can correctly predict the hydrolysis activity of a given ion, you can also consult Table 17.5 in the text.

Ex. 17.9 in the text illustrates calculating the pH of a salt solution. These calculations are really no different than those already illustrated for weak acids and bases in Exs. 17.6-8 in the text. *The only difference is that in the case of salts, ions are acting as weak acids and bases instead of neutral molecules.*

17.9 Polyprotic Acids and Bases

Polyprotic acids ionize in successive steps. For phosphoric acid, these steps (17aq, 17ar and 17as) and their respective K_a values are:

$H_3PO_4(aq) + H_2O(\ell) \rightleftharpoons H_3O^+(aq) + H_2PO_4^-(aq)$ $K_{a1} = 7.5 \times 10^{-5}$

$H_2PO_4^-(aq) + H_2O(\ell) \rightleftharpoons H_3O^+(aq) + HPO_4^{2-}(aq)$ $K_{a2} = 6.2 \times 10^{-8}$

$HPO_4^{2-}(aq) + H_2O(\ell) \rightleftharpoons H_3O^+(aq) + PO_4^{3-}(aq)$ $K_{a3} = 3.6 \times 10^{-13}$

Notice how *the successive K_a values get smaller, because it becomes more difficult to remove H^+ the more negative the charge on the acid.* Indeed, *each successive K_a is less than the previous one by a factor of about 10^4 to 10^6, and this is typical of oxoacids which are composed of H, O and a nonmetal. However, this is also means that each successive ionization step produces about a million times less hydronium ion than the step before it, so the pH of the solution depends primarily on the H_3O^+ generated by the first ionization step. Furthermore, $[A^{2-}] = K_{a2}$.*

The principles pertaining to polyprotic acids also apply to their respective conjugate anions. This means:

- each successive K_b is a factor of 10^4 to 10^6 less than the one before it, because the less negative the anion the lower the attraction for a H^+ ion from water and

- the pOH and pH depend primarily on the OH^- generated by the first step.

This is illustrated by the calculation of the pH of a carbonate solution in Ex. 17.10 in the text.

17.10 Molecular Structure, Bonding and Acid-Base Behavior

We can consider the dissociation of an acid HA to occur in steps to examine the correlation between molecular structure, chemical bonding and acid strength. These steps are:

1. H-A bond breaking with each atom retaining the electron it brought to the bond.

$$H-A \rightarrow H\cdot + \cdot A \qquad (17at)$$

2. Loss of an electron by H to form H^+.

$$H\cdot \rightarrow H^+ + e^- \qquad (17au)$$

3. Gain of electron by A to form A^-.

$$A\cdot + e^- \rightarrow A^- \qquad (17av)$$

4. Hydration of the H^+ and A^- ions to form aqueous ions.

$$H^+ + A^- \ -H_2O-> \ H_3O^+(aq) + A^-(aq) \qquad (17aw)$$

We can ignore Step 2, which is common to all acid dissociations, and furthermore ignore Step 4, because hydration can have about the same effect for analogous cases. Thus, we can focus on the H-A bond strength and the electron affinity of the fragment A when analyzing the strengths of similar acids. *In general, the weaker the H-A bond and the higher the electron affinity of the fragment A, the stronger the acid.* However, these factors often act in opposing directions, as in the hydrogen halide series.

In the case of the hydrogen halides, bond strength decreases with the increasing size of A from F to I whereas the electron affinities of the halogen atoms increase from F to Cl and then decrease with increasing size from Cl to I (8.6). So, according to the electron affinity of A data, HCl should be the strongest HA acid, and according to bond strength data, HI should be the strongest HA acid in the series. However, *there is a regular increase in acid strength from HF to HI that parallels the regular decrease in H-A bond strength and indicates the decrease in bond strength is the controlling factor in this series*. A similar pattern prevails for binary hydrides of elements of other Groups, so acid strength increases from H_2O to H_2Te in Group VIA. However, the combined electron affinity of fragment A and hydration energy for fragment A dominate going across a period, so acidity increases in the order CH_4, NH_3, H_2O and HF for binary hydrides of the nonmetals of the second period.

In the case of oxoacids composed of H, O and a nonmetal, the hydrogen atom that is ionized is always bonded to O, and the acidity is related to the composition of the rest of the molecule. For example, acid strength increases in the series HClO, $HClO_2$, $HClO_3$ and $HClO_4$, because there is an increasing number of O atoms attached to Cl that withdraw electron density from Cl and cause a shift in the electron density in the Cl-O-H bonds towards Cl via an **inductive effect** (attraction of electrons from adjacent bonds caused by a more electronegative atom). This makes the H atom more positive, more protonic, and the acid a stronger acid. *In general, the higher the number of O atoms attached to the central atom, the more protonic the hydrogen atom(s) and the stronger the acid*, so $HClO_3$ having two O atoms is expected to be stronger than H_3PO_4 having just one O atom. *As a matter of fact, all oxoacids having two or more O atoms attached to the central atom are expected to be strong acids*.

In the case of $HBrO_3$ and $HClO_3$, involving central atoms from the same group and the same number of O atoms, the higher electronegativity of Cl causes the H to be more positive and $HClO_3$ to be the stronger acid. *In general, the more electronegative the central atom, the stronger the acid when comparing acids of elements of the same group having the same number of O atoms.*

Lastly, it should be noted that only the H atom that is bonded to O in acetic acid, CH_3CO_2H is ionizable. This is because of the greater difference in electronegativity between O and H than C and H. *The H atom that is attached to the O atom is more positively charged, more protonic, and the O atom is better able to accommodate the negative charge left when the O-H bond breaks than C would be, if a C-H bond would break. This behavior is typical of carboxylic acids.*

17.11 The Lewis Concept of Acids and Bases

In the late 1930's, G.N. Lewis put forth a more general theory of acids and bases than the Bronsted theory. The **Lewis acid-base theory** is based on the sharing of an electron pair between an acid and a base, rather than the transfer of a proton. A **Lewis acid** is a substance that can accept a pair of electrons from another atom to form a new bond. A **Lewis base** is a substance that contains an atom with a lone pair of electrons that can be donated to another atom to form a new bond. *This can be represented as*

$$A + :B \rightarrow A:B \qquad\qquad (17ax)$$

where A represents the Lewis acid, B represents the Lewis base and AB represents the result, which is called an **adduct** or **complex**. The bond that is formed between the Lewis acid and base is a **coordinate covalent** bond (9.3). *The Lewis acid furnishes an empty orbital that overlaps with a filled orbital from the Lewis base.*

The Lewis theory is more general than the Bronsted theory. This means that every Bronsted acid is also a Lewis acid and that every Bronsted base is a Lewis base. Conversely, every Lewis base is a potential Bronsted base, because it has a lone pair of electrons with which to form a bond with (and thus accept) H^+, **but** not every Lewis acid is a Bronsted acid. Indeed, as we shall see, **not every Lewis acid contains a H atom.**

Cations as Lewis Acids

All metal cations are electron deficient and are therefore potential Lewis acids. Indeed, the formation of hydrated metal cations is an example of a Lewis acid-base reaction, and a consideration of the bonding in hydrated metal cations explains why they generally form acidic solutions. See p. 838 in the text.

Many metal ions also act as Lewis acids in forming complexes with the Lewis base ammonia. On the other hand, the amphoteric nature of a metal hydroxide involves the metal hydroxide acting separately as a Lewis acid and a Bronsted base. Consider, Al(OH)$_3$.

- In its reaction with bases, $Al(OH)_3$ behaves as a Lewis acid: $Al(OH)_3(s) + OH^-(aq) \rightarrow Al(OH)_4^-(aq)$. (17ay)

- In its reaction with acids, $Al(OH)_3$ behaves as a Bronsted base: $Al(OH)_3(s) + 3 H_3O^+(aq) \rightarrow Al^{3+}(aq) + 6 H_2O(\ell)$. (17az)

Molecules as Lewis Acids

The Lewis theory accounts nicely for the fact that oxides of nonmetals behave as acids. The highly electronegative oxygen atoms withdraw electrons from the central atom causing it to be somewhat positively charged and subject to attack by the Lewis bases OH^- or H_2O, as illustrated on p. 840 in the text.

17.12 Key Expressions

Autoionization of water expression:

$$H_2O(\ell) + H_2O(\ell) \rightleftharpoons H_3O^+(aq) + OH^-(aq)$$

Bronsted acid-base equation expression:

$$\text{Acid 1 + Base 2} \rightleftharpoons \text{Acid 2 + Base 1}$$

where Acid 1 and Base 1 are a conjugate acid-base pair and Base 2 and Acid 2 are a conjugate acid-base pair.

Weak acid ionization constant expression:

$$K_a = \frac{[H_3O^+][A^-]}{[HA]}$$

Weak base ionization constant expression:

$$K_b = \frac{[BH^+][OH^-]}{[B]}$$

Water ionization constant expression:

$$K_w = [H_3O^+][OH^-]$$

$$K_w = [H_3O^+][OH^-] = 1.0 \times 10^{-14} \text{ at } 25°C$$

Water ionization constant, weak acid ionization constant and conjugate base ionization constant relation expression:

$$K_a \cdot K_b = K_w$$

$$K_a \cdot K_b = K_w = 1.0 \times 10^{-14} \text{ at } 25°C$$

pH expressions:

$$pH = -\log[H_3O^+]$$

$$[H_3O^+] = antilog(-pH) = 10^{-pH}$$

pOH expressions:

$$pOH = -\log[OH^-]$$

$$[OH^-] = antilog(-pOH) = 10^{-pOH}$$

pH and pOH relation expression:

$$pH + pOH = 14 \text{ at } 25°C$$

percent ionization expression:

$$\%ionized = \frac{quantity\ ionized}{initial\ concentration} \cdot 100\%$$

$$= \frac{x}{[HA]_o} \cdot 100\% \text{ } \underline{\textbf{or}} \text{ } \frac{x}{[B]_o} \cdot 100\%$$

17.13 Hints for Answering Questions and Solving Problems

Identifying strong and weak acids and bases:

There are only a few strong acids and bases, **so the easiest way to differentiate between weak and strong acids and bases is to memorize the list of strong acids and bases. They are:**

Strong Acids:

Nitric acid:	HNO_3
Sulfuric acid:	H_2SO_4 (for loss of first H^+ only)
Halic acids:	$HClO_3$, $HBrO_3$
Perhalic acids:	$HClO_4$, $HBrO_4$
Hydrohalic acids:	HCl, HBr and HI

Strong Bases:

1A hydroxides:	LiOH, NaOH, KOH, RbOH, CsOH
Soluble 2A hydroxides:	$Ca(OH)_2$, $Sr(OH)_2$, $Ba(OH)_2$*
Stronger than OH^-:	S^{2-}, $C_2H_5O^-$, NH_2^-, H^-, CH_3^-

*See p. 41 concerning the molar solubility limits for these $M(OH)_2$ compounds acting as strong bases. ✶ Try preparing a "flash card" listing these strong acids and bases and memorizing them.

Working with strong and weak acids and bases:

Recall that **all strong acids react 100% in ionization reactions with H_2O to give H_3O^+,**

$$HA(aq) + H_2O(\ell) \rightarrow H_3O^+(aq) + A^-(aq)$$

the strongest acid that can exist in water. <u>**So equilibrium expressions are not used when working with strong acids and K_a values are not given for strong acids**</u> (See Table 17.4.).

Also recall that **the strong bases are the soluble ionic hydroxides which are considered to produce OH^- in 100% ionic dissociation reactions in water**

$$MOH(s) \xrightarrow{H_2O} M^+(aq) + OH^-(aq)$$
$$\approx 100\% \text{ reaction}$$

and

$$M(OH)_2(s) \xrightarrow{H_2O} M^{2+}(aq) + 2OH^-(aq)$$
$$\approx 100\% \text{ reaction}$$

or anions which react 100% in ionization reactions with H_2O to give OH^-,

$$B^-(aq) + H_2O(\ell) \rightarrow BH(aq) + OH^-(aq)$$

or

$$S^{2-}(aq) + H_2O(\ell) \rightarrow HS^-(aq) + OH^-(aq)$$

the strongest base that can exist in water. <u>So, equilibrium expressions are not used when working with strong bases, and K_b values are not given for the strong bases, except S^{2-}</u> (See Table 17.4.).

On the other hand, **weak acids and bases only partially react with water, so equilibrium expressions are used with weak acids and bases and K_a and K_b values are given in Tables like 17.4 for weak acids and bases.** Recall Expressions 17t-w pertaining to weak acid and base equilibria.

Conducting equilibrium calculations for weak acids and bases:

A systematic method for calculating equilibrium concentrations in solutions of weak acids and bases was described on pp. 235 and 236. However, a shorter method can be used whenever the approximation method is valid. It can be shown that:

$$\text{when } [HA]_o \geq 100 \cdot K_a, \ x = [H_3O^+] = [A^-] = \sqrt{K_a \cdot [HA]_o}$$

and

$$\text{when } [B]_o \geq 100 \cdot K_b, \ x = [BH^+] = [OH^-] = \sqrt{K_b \cdot [B]_o}$$

Using correct numbers of significant figures for pH values from $[H_3O^+]$ and $[H_3O^+]$ values from pH:

Recall that two significant figures for $[H_3O^+]$ leads to two decimal places for pH and vice-versa. Numbers to the left of the decimal place in logarithmic numbers are related to the power of ten for the number when written in scientific notation.

Recalling Lewis acid-base definitions:

H^+ is a Bronsted acid and a Lewis acid and has no electrons. Thus, Lewis acids must be electron-pair acceptors, and Lewis bases must be electron-pair donors.

CHAPTER 18

PRINCIPLES OF REACTIVITY:
REACTIONS BETWEEN ACIDS AND BASES

Acid-base reactions occur within us and around us and represent some of the most important reactions known. For example, the manufacture of the fertilizer ammonium phosphate, the action of acid rain on limestone statutes and buildings and on the components of the soil and the buffering capacity of our blood are all based on acid-base reactions. This chapter examines the driving forces for the various types of acid-base reactions, the nature of buffer solutions and acid-base titrations, which can be used for chemical analysis.

18.1 Acid-Base Reactions

The predominant direction of acid-base reactions is always towards forming the weaker acid-base pair. This occurs because the stronger acid has a greater tendency to react with the stronger base than the weaker acid has to react with the weaker base. *Thus, tables showing relative acid-base strengths can be used to predict the outcomes of acid-base reactions (17.3).* For example, we can predict that acetic acid will react to a significant extent with ammonia in water

$$CH_3CO_2H(aq) + NH_3(aq) \rightleftharpoons NH_4^+(aq) + CH_3CO_2^-(aq) \qquad (18a)$$

because, according to Table 17.4, CH_3CO_2H is a stronger acid than NH_4^+ and NH_3 is a stronger base than $CH_3CO_2^-$.

We can also calculate the equilibrium constant for the reaction of acetic acid with ammonia (18a) to verify our prediction, because it can be considered to be the sum of three equations whose equilibrium constants are known (16.1).

Ionization of the acid:

$$CH_3CO_2H(aq) + H_2O(\ell) \rightleftharpoons H_3O^+(aq) + CH_3CO_2^-(aq) \qquad (18b)$$

$$K_a = 1.8 \times 10^{-5}$$

Ionization of the base:

$$NH_3(aq) + H_2O(\ell) \rightleftharpoons NH_4^+(aq) + OH^-(aq) \qquad (18c)$$

$$K_b = 1.8 \times 10^{-5}$$

Union of hydronium ion and hydroxide ion:

$$H_3O^+(aq) + OH^-(aq) \rightleftharpoons 2\ H_2O(\ell) \qquad (18d)$$

$$K = \frac{1}{K_w} = 1.0 \times 10^{14}$$

Net reaction:

$$CH_3CO_2H(aq) + NH_3(aq) \rightleftharpoons NH_4^+(aq) + CH_3CO_2^-(aq) \qquad (18e)$$

$$K_{net} = \frac{K_a K_b}{K_w} = 3.2 \times 10^4$$

The value of K_{net} is fairly large indicating the reaction is definitely product-favored. Although acetic acid is a weak acid and ammonia is a weak base, the H_3O^+ and OH^- ions they produce are very strong and are consumed by the water formation reaction with its large equilibrium constant of 1.0×10^{14}. Thus, as suggested by LeChatelier's principle, the removal of the H_3O^+ and OH^- ions by water formation causes the acid and base ionization reactions to proceed much further to the right than either would go if it occurred alone. The reaction of a weak acid with a weak base is therefore driven by the formation of the weak electrolyte water, though water does not appear in the equation for the net reaction.

The reaction of a weak acid with a weak base is just one of four possible types of acid-base reactions that can occur in aqueous solution. These four are:

Type	Strength of Acid	Strength of base
1	strong	strong
2	strong	weak
3	weak	strong
4	weak	weak

and we will consider the outcome of each type in detail below.

The Reaction of a Strong Acid with a Strong Base

Strong acids and bases are effectively 100% ionized in solution (17.4). Thus, for the reaction of HCl with NaOH, we can write

$$H_3O^+(aq) + Cl^-(aq) + Na^+(aq) + OH^-(aq) \rightarrow \qquad (18f)$$
$$Na^+(aq) + Cl^-(aq) + 2\ H_2O(\ell)$$

which leads to the net ionic equation

$$H_3O^+(aq) + OH^-(aq) \rightarrow 2\ H_2O(\ell) \tag{18g}$$

$$K_{net} = \frac{1}{K_w} = 1.0 \times 10^{14}$$

The net ionic equation for the reaction of any strong acid with a strong metal hydroxide base involves the union of hydronium ions and hydroxide ions to form water, Expression 18g. The large value of K for this reaction ensures the reaction goes to completion, so the mixing of equal moles of HCl and NaOH yields a solution of NaCl in water. Because the Na^+ and Cl^- ions of NaCl come from a strong base and strong acid and do not react appreciably with water (17.8), the solution is a **neutral solution**, and the reaction is called a **neutralization reaction.** *The mixing of strong acids and bases in amounts yielding equal moles of hydronium and hydroxide ions always yields a neutral solution.*

The Reaction of a Strong Acid with a Weak Base

The reaction of HCl with NH_3 is a classic example of the reaction of a strong acid with a weak base. *The $H_3O^+(aq)$ produced by the 100% ionization of HCl reacts with the OH^- produced by the weak base ionization of NH_3.*

$$NH_3(aq) + H_2O(\ell) \rightleftharpoons NH_4^+(aq) + OH^-(aq) \quad K_b = 1.8 \times 10^{-5} \tag{18h}$$

$$H_3O^+(aq) + OH^-(aq) \rightarrow 2\ H_2O(\ell) \qquad K = \frac{1}{K_w} = 1.0 \times 10^{14} \tag{18i}$$

$$\overline{H_3O^+(aq) + NH_3(aq) \rightleftharpoons NH_4^+(aq) + H_2O(\ell)} \tag{18j}$$

$$K_{net} = \frac{K_b}{K_w} = 1.8 \times 10^9$$

Again the large value of the equilibrium constant indicates the reaction proceeds to completion, so the mixing of equal numbers of moles of HCl and NH_3 yields a solution of NH_4Cl in water, a salt that is 100% dissociated in water. *The NH_4^+ is the conjugate acid of a weak base and therefore acts as a weak acid whereas the Cl^- is the conjugate base of a strong acid and does not react appreciably with water (17.8), so the solution is acidic.*

$$NH_4^+(aq) + H_2O(\ell) \rightleftharpoons H_3O^+(aq) + NH_3(aq) \tag{18k}$$

$$K_a \text{ for } NH_4^+ \text{ from Table 17.4} = 5.6 \times 10^{-10}$$

The reaction of a strong acid with a weak base always gives an acidic solution at the equivalence point, the pH of which depends on the concentration of the cation from the salt formed

and its K_a. Ex. 18.1 in the text illustrates the procedure for calculating this pH. **Be sure to notice the need to do stoichiometric calculations to determine the moles of acid and base combined (mol acid = $L_{acid} \cdot M_{acid}$ and mol base = $L_{base} \cdot M_{base}$) and thus the moles of salt formed and then to divide the moles of salt formed by the total volume to get the molar concentrations of the salt and of the ions from the salt at the equivalence point.** ✱ Try preparing a "flash card" summarizing the procedure for calculating the pH at the equivalence point of a strong acid-weak base reaction and memorizing it. Also try practicing doing this type of calculation until you are comfortable doing them and can do them quickly.

The Reaction of a Weak Acid with a Strong Base

Consider the reaction of formic acid, HCO_2H, the weak acid that causes the sting associated with ant bites, with the strong base NaOH. **The H_3O^+ that is produced by the ionization of the HCO_2H reacts with the OH^- produced by the 100% dissociation of NaOH.**

$$HCO_2H(aq) + H_2O(\ell) \rightleftharpoons H_3O^+(aq) + HCO_2^-(aq) \quad K_a = 1.8 \times 10^{-4} \quad (18\ell)$$

$$H_3O^+(aq) + OH^-(aq) \rightarrow 2\ H_2O(\ell) \qquad K = \frac{1}{K_w} = 1.0 \times 10^{14} \quad (18m)$$

$$\overline{HCO_2H(aq) + OH^-(aq) \rightarrow H_2O(\ell) + HCO_2^-(aq)} \qquad\qquad (18n)$$

$$K_{net} = \frac{K_a}{K_w} = 1.8 \times 10^{10}$$

Once again the large value of the equilibrium constant indicates the reaction proceeds to completion, so the mixing of equal numbers of moles of HCO_2H and NaOH yields a solution of $NaHCO_2$ and water, a salt that is 100% dissociated in water. The Na^+ is the cation of a strong base and does not react appreciably with water whereas the HCO_2^- is the conjugate base of a weak acid and therefore acts as a weak base (17.8), so the solution is basic.

$$HCO_2^-(aq) + H_2O(\ell) \rightarrow HCO_2H(aq) + OH^-(aq) \qquad (18o)$$

$$K_b \text{ for } HCO_2^- \text{ from Table 17.4} = 5.6 \times 10^{-11}$$

The reaction of a weak acid with a strong base always gives a basic solution at the equivalence point, the pH of which depends on the concentration of the anion from the salt formed and its K_b. Ex. 18.2 in the text illustrates the procedure for calculating this pH. **Once again be sure to notice the need to do stoichiometric calculations to determine the moles of acid and base combined (mol acid = $L_{acid} \cdot M_{acid}$ and mol base = $L_{base} \cdot M_{base}$) and thus the moles of salt formed and then to divide the moles of salt formed by the total volume to get the molar concentra-**

tions of the salt and of the ions from the salt at the equiva-lence point. ✱ Try preparing a "flash card" summarizing the procedure for calculating the pH at the equivalence point of a strong acid-weak base reaction and memorizing it. Also try practicing doing this type of calculation until you are comfortable doing them and can do them quickly.

The Reaction of a Weak Acid with a Weak Base

We have already considered the reaction of CH_3CO_2H with NH_3 to yield a solution of $NH_4CH_3CO_2$ and water, a salt that is 100% dissociated in water. The NH_4^+ cation is the conjugate of the weak base NH_3 and therefore acts as a weak acid

$$NH_4^+(aq) + H_2O(\ell) \rightleftharpoons H_3O^+(aq) + NH_3(aq) \qquad (18p)$$

$$K_a = 1.8 \times 10^{-5}$$

whereas the $CH_3CO_2^-$ anion is the conjugate base of the weak acid CH_3CO_2H and therefore acts as a weak base

$$CH_3CO_2^-(aq) + H_2O(\ell) \rightleftharpoons CH_3CO_2H(aq) + OH^-(aq) \qquad (18q)$$

$$K_b = 1.8 \times 10^{-5}$$

Here the concentrations of NH_4^+ and $CH_3CO_2^-$ from $NH_4CH_3CO_2$ are equal and the value of K_a for NH_4^+ is equal to the value of K_b for $CH_3CO_2^-$, so the solution is expected to be neutral at the equivalence point in the reaction of CH_3CO_2H with NH_3. *In general, the pH of the solution depends on the relative values of K_a for the cation and K_b for the anion of the salt formed.*

A Summary of Acid-Base Reactions

• When the reaction involves a strong acid and a strong base, the solution is neutral at the equivalence point, and the equilibrium constant for the net reaction is:

$$K_{net} = 1/K_w = 1.0 \times 10^{14} \qquad (18r)$$

• When the reaction involves a strong acid and a weak base, the solution is acidic at the equivalence point, and the equilibrium constant for the net reaction is given by:

$$K_{net} = K_b/K_w \qquad (18s)$$

where K_b is the ionization constant for the weak base that is reacting.

• When the reaction involves a weak acid and a strong base, the solution is basic at the equivalence point and the equilibrium constant for the net reaction is given by:

$$K_{net} = K_a/K_w \tag{18t}$$

where K_a is the ionization constant for the weak acid that is reacting.

• When the reaction involves a weak acid and a weak base, the solution can be acidic, neutral or basic at the equivalence point, and the equilibrium constant for the net reaction is given by:

$$K_{net} = \frac{K_a K_b}{K_w} \tag{18u}$$

where K_a and K_b are the ionization constants for the weak acid and weak base reacting, respectively.

The pH at the equivalence point of a weak acid-weak base reaction depends on the K_a of the weak acid cation formed and the K_b of the weak base anion formed. The three possibilities are (17.8):

K_a cation > K_b anion   acidic solution
K_a cation = K_b anion   neutral solution
K_a cation < K_b anion   basic solution

In other words, the pH at the equivalence point in an acid-base reaction depends on the nature of the cation and anion formed by the acid-base reaction, and the outcome is the same as was predicted for salt hydrolysis in 17.8. Thus, you should review the summary table that was given on p. 237 concerning the acid-base nature of salt solutions as needed. ✶ Try preparing a separate "flash card" for each of the four types of acid-base reactions stating the general form of K_{net}, the specific identity of K_a or K_b terms appearing in K_{net} and the pH to be expected at the equivalence point and memorize them. Also be prepared to rationalize the pH to be expected at the equivalence point in terms of salt hydrolysis and to use chemical equations to illustrate your answer.

18.2 The Common Ion Effect

The Common Ion Effect

The **common ion effect** refers to the influence of an ion that is in solution and is "common" to a chemical equilibrium, that is to the influence of an ion that is in solution and part of a chemical equilibrium. **The source of the common ion can be a strong acid (a source of H_3O^+), strong base (a source of OH^-) or a soluble ionic salt, including a soluble ionic salt that is being formed during the reaction of a weak acid or weak base. The common ion is usually a product ion which opposes the ionization of the weak acid or base in accord with LeChatelier's principle. The concentration of the common ion becomes an initial concentration in the ICE table for solving for equilibrium concentrations**, as illustrated in Ex. 18.3 in the text.

18.3 Buffer Solutions

A **buffer solution** resists change in pH when small amounts of strong base or strong acid are added. **A buffer solution must therefore contain an acid that can react with added OH^- ions and a base that can react with added H_3O^+ ions. In addition, the acid and base must not react with each other. Thus, a buffer is usually composed of roughly equal quantities of the members of a conjugate pair: a weak acid and its weak conjugate base or a weak base and its weak conjugate acid.**

Ex. 18.4 in the text illustrates how to calculate the pH of a buffer when the concentrations of the components are known, as well the change in pH that occurs when a small amount of HCl is added to 1.0 L of water and when the same amount of HCl is added to 1.0 L of an acetic acid/sodium acetate buffer. **Notice that added strong acid reacts with the conjugate base component of a buffer whereas added strong base reacts with the acid component of a buffer. Also notice how it is useful to make a table showing the moles of all species present before the strong acid reacts with the conjugate base and to calculate the concentrations of the buffer components following 100% reaction with the added strong acid or strong base.** ✴ Try preparing a "flash card" summarizing the steps to be used to calculate the pH of a buffer solution following the addition of strong acid or strong base and memorizing it. Also try practicing doing these types of calculations until you are comfortable doing them and can do them quickly.

General Expressions for Buffer Solutions

Buffer Solutions Based on Weak Acids:

Rearranging the K_a expression for a weak acid, HA, gives the following general expression for the concentration of hydronium ion in a buffer solution in terms of the concentration of the weak acid, HA, and its conjugate base, A^-:

$$[H_3O^+] = \frac{[\text{acid}]}{[\text{conjugate base}]} \cdot K_a \qquad (18v, 18.1)$$

Because the equilibrium concentrations of the acid and its conjugate base are virtually equal to their initial concentrations in a buffer solution, the change in concentration, x, can be neglected and the initial concentrations can be substituted directly into Expression 18v to calculate the hydronium ion concentration. (Recall the solution to Ex. 18.3b on p. 857 in the text where the change in concentration was neglected with respect to both the initial acid and conjugate base concentrations to solve for the hydronium ion concentration, and note that a buffer solution is just a special case of the common ion effect. This is why similar approximations could be made in the solution to Ex. 18.4 on p. 861 in the text.)

Buffer Solutions Based on Weak Bases:

Rearranging the K_b expression for a weak base, B, gives the following general expression for the concentration of hydroxide ion in a buffer solution in terms of the concentration of the weak base, B, and its conjugate acid, BH^+:

$$[OH^-] = \frac{[\text{base}]}{[\text{conjugate acid}]} \cdot K_b \qquad (18w, 18.2)$$

Once again, changes in concentrations can be neglected mathematically and initial concentrations can be substituted directly into Expression 18w.

Henderson-Hasselbach Equation for Buffers Based on Weak Acids

The general buffer expression for a weak acid (18v) can be mathematically converted to

$$pH = pK_a + \log\frac{[\text{conjugate base}]}{[\text{acid}]} \qquad (18x, 18.3)$$

This equation, known as the **Henderson-Hasselbach equation,** *is valid when the ratio of [conjugate base]/[acid] is no larger than $\approx 10/1$ and no smaller than $\approx 1/10$, because these are the ranges in which there can be buffering action.* However, *the best buffering action comes when the ratio is approximately 1, and the pH is therefore approximately equal to the pK_a of the acid.*

Notice that the form of Expression 18x tells us that **the pH of a solution of a weak acid and its conjugate base is controlled primarily by the strength of the acid, as expressed by its pK_a.** *The "fine control" of the pH is then given by the relative amounts of acid and conjugate base. If*

• the amount of acid is less than the amount of conjugate base, the pH is less than the pK_a, because $\log([\text{conjugate base}]/[\text{acid}])$ is negative.

• the amount of acid is equal to the amount of conjugate base, the pH is equal to the pK_a, because the log of 1 is zero.

• the amount of acid is greater than the amount of conjugate base, the pH is greater than the pK_a, because $\log([\text{conjugate base}]/[\text{acid}])$ is positive.

However, the [conjugate base]/[acid] ratio range from 10/1 to 1/10 for buffering action, limits the pH range for the buffering action of any given weak acid to

$$\text{pH range for buffering} = pK_a \pm 1 \qquad (18y)$$

✶ Try preparing a "flash card" containing the Henderson-Hasselbach equation and a summary of the three possibilities for the ratio [conjugate base]/[acid] and the resulting pH values relative to the pK_a value and memorizing it. Also make sure you understand the reasoning for the resulting relative pH values.

Ex. 18.5 in the text illustrates the use of the Henderson-Hasselbach equation to calculate the pH of a buffer system. Notice that *it is possible to use the number of moles of conjugate base provided by the salt of the weak acid and the number of moles of acid present in place of their respective molar concentrations when using Expression 18v or 18x.* This is *because both substances necessarily occupy the same volume, and the mole ratio is necessarily identical to the molar concentration ratio.*

Preparing Buffer Solutions

There are two requirements for a buffer solution:

• it must control the pH at the desired value, so Expression 18y tells us the weak acid selected must have a pK_a that is within one unit of the desired pH.

• it must have sufficient acid and base forms present to react with added strong acid or strong base and resist change in pH, so buffer solutions usually have 0.1 to 1.0 M concentrations of weak acid and conjugate base or weak base and conjugate acid.

However, it should be recognized that *any buffer loses its buffering capacity, if too much strong acid or strong base is added.*

Ex. 18.6 illustrates the procedure for selecting an acid/conjugate base pair to be used to prepare a buffer of a given pH and for calculating the [conjugate base]/[acid] ratio required to obtain that pH. Notice two things:

• The concentration ratio required also gives the mole ratio required and any combination of masses giving this mole ratio can be used.

• Any volume can be prepared with the desired mole ratio, provided the molar concentrations of the components exceed 0.1 M to ensure sufficient buffering capacity.

This also illustrates another important property of buffer solutions. *The mole ratio is more important than the volume in determining the pH. Thus, diluting a buffer solution does not change its pH, because it does not change the conjugate base/acid mole ratio.*

18.4 Acid-Base Titration Curves

Acid-base titration curves involve plots of pH versus mL of reagent (acid or base) added from a buret and can be used to determine the equivalence point of the reaction or select an appropriate acid-base indicator for the reaction. The titration curve in Fig. 18.5 in the text for HCl with NaOH is typical of titration curves for titrations of strong acids with strong bases, and the titration curve in Fig. 18.6 in the text for CH_3CO_2H with NaOH is typical of titration curves for titrations of weak acids with strong bases.

There are several important differences between the titration curves for the titration of a strong acid with a strong base and the titration of a weak acid with a strong base. These are:

• The initial pH of the strong acid solution is lower than that of the weak acid of equal concentration, because the strong acid is 100% ionized and therefore yields a higher $[H_3O^+]$ than the weak acid which ionizes only slightly.

• The pH of the strong acid solution increases steadily at first whereas the pH of the weak acid solution shows an initial jump followed by a buffer region in which it changes more gradually.

• The pH at the half-neutralization point of the weak acid is equal to the pK_a of the acid, because one-half of the acid has been converted to its conjugate base and the [conjugate base]/[acid] ratio is 1. There is no correspond-ing point for the titration of a strong acid with a strong base, because there is little, if any, buffering action in such systems.

• The pH at the equivalence point of the strong acid with strong base reaction is 7 whereas that of the weak acid with strong base reaction is >7 due to hydrolysis of the conjugate weak base anion that is formed (18.1).

• The pH rise near the equivalence point is larger for the strong acid than it is for the weak acid. In general, **the stronger the acid the greater the rise in pH near the equivalence point**. Indeed, in the case of weak acid with weak base titrations, there is a buffer region before and after the equivalence point and little change in pH at the equivalence point, so the equivalence point cannot be determined using ordinary means.

The pOH at the half-neutralization point of the titration of a weak base with a strong acid is similarly equal to the pK_b of the base, so titration curves can also be used to determine K_a and K_b values.

Calculations for Titration Curves:

There are four regions of interest for titration curves: the starting point, the region between the starting point and the equivalence point, the equivalence point and the region beyond the equivalence. We will consider the procedure for calculating the pH in each of these regions for strong acid-strong base, weak acid-strong base and weak base-strong acid titrations.

Strong Acid-Strong Base

• The pH at the starting point (0.00 mL base added) is due to $[H_3O^+]$ formed by 100% ionization of the strong acid. The calculation is like that for Exercise. 17.7 in the text.

• The pH at any point between the starting point and the equivalence point is due to the excess $[H_3O^+]$ which can be calculated using

$$[H_3O^+] = \frac{\text{original moles acid - moles base added}}{\text{volume acid (L) + volume base added (L)}} \quad (18z)$$

where mol acid = $L_{acid} \cdot M_{acid}$ and mol base = $L_{base} \cdot M_{base}$. The calculation is like that in Steps 2 and 3 in Ex. 18.7 in the text.

• The pH at the equivalence point is 7 because the cation of the salt formed comes from a strong base and does not react appreciably with water and the anion of the salt formed comes from a strong acid and does not react appreciably with water (18.1).

• The pH at any point beyond the equivalence point is due to $[OH^-]$ formed by 100% dissociation of the excess strong base.

$[OH^-]$ = <u>moles base added - original moles acid</u>   (18ab)
        volume acid (L) + volume base added (L)

Weak Acid-Strong Base

• The pH at the starting point (0.00 mL base added) is due to the $[H_3O^+]$ formed by ionization of the weak acid. The calculation is like that found in Ex. 17.6 in the text.

• The pH at any point between the starting point and the equivalence point is due to the HA/A^- buffer that is created as some of the weak acid, HA, is converted to its conjugate base, A^-, by reaction with the strong base. The pH can be calculated by substituting the moles of acid remaining and the moles of conjugate base formed into either Expression 18v or 18x. These quantities can be calculated using,

mol acid   = original mol acid - mol base added (18ac)
remaining

and

mol conjugate base formed = mol base added   (18ad)

where mol acid = $L_{acid} \cdot M_{acid}$ and mol base = $L_{base} \cdot M_{base}$. The calculation is like that found in Ex. 18.8 in the text.

• The pH at the equivalence point is due to hydrolysis of the conjugate weak base anion, A^-, that is formed during the titration.

$$A^-(aq) + H_2O(\ell) \rightleftharpoons HA(aq) + OH^-(aq)$$ (18ae)

The calculation is like that found in Ex. 18.2 in the text.

• The pH at any point beyond the equivalence point is due to $[OH^-]$ formed by 100% dissociation of the excess strong base.

$$[\text{OH}^-] = \frac{\text{moles base added - original moles acid}}{\text{volume acid (L) + volume base added (L)}} \quad \text{(18af)}$$

Weak Diprotic Acid-Strong Base

Oxalic acid, $H_2C_2O_4$, is weak diprotic acid which gives rhubarb its tart taste. The ionization expressions for oxalic acid are:

$$H_2C_2O_4(aq) + H_2O(\ell) \rightleftharpoons H_3O^+(aq) + HC_2O_4^-(aq) \quad \text{(18ag)}$$

$$K_{a1} = 5.9 \times 10^{-2}$$

$$HC_2O_4^-(aq) + H_2O(\ell) \rightleftharpoons H_3O^+(aq) + C_2O_4^{2-}(aq) \quad \text{(18ah)}$$

$$K_{a2} = 6.4 \times 10^{-5}$$

The relatively large value for K_{a1} suggests there should be a relatively large increase in pH near the first equivalence point, but the titration curve shown in Fig. 18.7 in the text indicates this is not the case. The reason for this is that there is buffering action on both sides of the first equivalence point. *There is buffering due to the $H_2C_2O_4/HC_2O_4^-$ conjugate pair prior to the first equivalence point and buffering due to the $HC_2O_4^-/C_2O_4^{2-}$ conjugate pair following the first equivalence point. Thus, the species in solution resist a rapid change in pH at the first equivalence point, and a larger increase in pH occurs at the second equivalence point, though $K_{a2} < K_{a1}$.*

The pH at the second equivalence point is due to hydrolysis of the $C_2O_4^{2-}$ ion that is formed following the first equivalence point.

$$C_2O_4^{2-}(aq) + H_2O(\ell) \rightleftharpoons HC_2O_4^-(aq) + OH^-(aq) \quad \text{(18ai)}$$

$$K_b = K_w/K_{a2} = 1.6 \times 10\text{-}10$$

This is why the pH at the second equivalence point is >7.

Weak Base-Strong Acid

• The pH at the starting point (0.00 mL acid added) is due to the [OH⁻] formed by ionization of the weak base. The calculation is like that found in Ex. 17.8 in the text.

• The pH at any point between the starting point and the equivalence point is due to the B/BH⁺ buffer that is created as some of the weak base, B, is converted to its conjugate acid, BH⁺, by reaction with the strong acid. The pH can be calculated by substituting the moles of base remaining and the moles of conjugate acid formed into Expression 18w. These quantities can be calculated using,

    mol base   = original mol base - mol acid added (18aj)
    remaining

and

    mol conjugate acid formed = mol acid added   (18ak

where mol base = $L_{base} \cdot M_{base}$ and mol acid = $L_{acid} \cdot M_{acid}$. The calculation is similar to that found in Ex. 18.8 in the text but uses the expression for buffers based on weak bases (18w, 18.3).

• The pH at the equivalence point is due to hydrolysis of the conjugate weak acid cation, BH^+, that is formed during the titration.

$$BH^+(aq) + H_2O(\ell) \rightleftharpoons H_3O^+(aq) + B(aq) \qquad (18a\ell)$$

The calculation is like that found in Ex. 18.1 in the text.

• The pH at any point beyond the equivalence point is due to $[H_3O^+]$ formed by 100% ionization of the excess strong acid.

$$[H_3O^+] = \frac{\text{moles acid added - original moles base}}{\text{volume base (L) + volume acid added (L)}} \qquad (18am)$$

18.5 Acid-Base Indicators

The goal of an acid-base titration is to add an amount of acid or base that is stoichiometrically equivalent to the amount of base or acid being titrated. This occurs at the **equivalence point** of the reaction. *Titration curves can be used to determine when the equivalence point is reached, because the equivalence point corresponds to the midpoint of the rapid pH rise. However, acid-base indicators are commonly used to estimate when the equivalence point has been reached.* An **acid-base indicator** is a substance that can exist in conjugate acid and base forms having different colors and is known to change colors in a given pH range (17.6). We say the color change occurs at the **end point** of the titration, because that is the point at which we end the titration, and *we try to select an indicator that will make the volume difference between the end point and equivalence point negligible.*

An acid-base indicator is usually an organic dye that is a weak acid that can be represented by HInd, so

$$HInd(aq) + H_2O(\ell) \rightleftharpoons H_3O^+(aq) + Ind^-(aq) \qquad (18an)$$

and

$$K_a \text{ HInd} = \frac{[H_3O^+][Ind^-]}{[HInd]} \qquad (18ao)$$

or

$$[H_3O^+] = \frac{[HInd]}{[Ind^-]} \cdot K_a \text{ HInd} \qquad (18ap)$$

which tells us the [HInd]/[Ind⁻] is controlled by the [H₃O⁺] and pH of the solution. This is important because the [HInd]/[Ind⁻] ratio determines the color of the solution. When [HInd]/[Ind⁻] ≈ 10/1, the color of HInd predominates, and when [HInd]/[Ind⁻] ≈ 1/10, the color of Ind⁻ predominates. This means the color change occurs over a hydronium ion concentration range of about 100, which corresponds to 2 pH units, and the pH range of an indicator is given by

$$\text{pH range of HInd} = pK_a \text{ HInd} \pm 1 \qquad (18aq)$$

This, however, isn't much of a problem, because the pH change at the equivalence point is usually large (See Figs. 18.6 and 7 in the text.) and there is a variety of indicators available having differing pH ranges (See Fig. 17.7 in the text.). For example, phenolphthalein, which changes from colorless to pink in the pH range 8 to 10 is suitable for use with strong acid-strong base, strong base-strong acid and weak acid-strong base titrations, because the pH at the equivalence point for the first two types is 7 and that for the latter is >7 due to anion hydrolysis. However, methyl red, which changes in the pH range 4.2 to 6.2 is better suited for use with a weak base-strong acid titration, because the pH at the equivalence point is <7 due to cation hydrolysis.

18.6 Key Expressions

K_{net} *expressions for acid-base reactions:*

Strong acid-strong base: $K_{net} = \dfrac{1}{K_w} = 1.0 \times 10^{14}$

Strong acid-weak base: $K_{net} = K_b/K_w$

where K_b is for the weak base that is reacting.

Weak acid-strong base: $K_{net} = K_a/K_w$

where K_a is for the weak acid that is reacting.

Weak acid-weak base: $K_{net} = \dfrac{K_a K_b}{K_w}$

where K_a and K_b are for the weak acid and weak base reacting, respectively.

Weak acid buffer expressions:

$$[H_3O^+] = \frac{[acid]}{[conjugate\ base]} \cdot K_a$$

$$pH = pK_a + \log\frac{[conjugate\ base]}{[acid]}$$

$$pH\ range\ for\ buffering = pK_a \pm 1$$

Weak base buffer expression:

$$[OH^-] = \frac{[base]}{[conjugate\ acid]} \cdot K_b$$

pH range of acid-base indicator expression:

$$pH\ range\ of\ HInd = pK_a\ HInd \pm 1$$

18.7 Hints for Answering Questions and Solving Problems

Predicting the pH at the equivalence point of an acid-base reaction/titration:

The following summary can be used to predict the pH at the equivalence point of an acid-base reaction:

Acid	Base	pH at Equivalence Point
Strong	Strong	= 7 (neutral)
Strong	Weak	<7 (acidic due to cation hydrolysis)
Weak	Strong	>7 (basic due to anion hydrolysis)
Weak	Weak	Depends on K_a of cation formed and K_b of anion formed.

Recall that the cation of the salt present at the equivalence point comes from the base and the anion comes from the acid, so this table is in agreement with that given on p. 237 for salt hydrolysis. * Try preparing a "flash card" summarizing this information and memorizing it.

Calculating the pH at the equivalence point of an acid-base reaction/titration:

No pH calculation is necessary for the equivalence points of titrations of strong acids with strong bases or strong bases with strong acids, because the ions of the salts formed by these combinations do not react with water appreciably. The pH is therefore predicted to be 7, assuming the pH of the water used to prepare the solutions is 7.

The procedure for calculating the pH at the equivalence point for the titration of a weak base with a strong acid is summarized on p. 248 and illustrated in Ex. 18.1 in the text.

The procedure for calculating the pH at the equivalence point for the titration of a weak acid with a strong base is summarized on p. 248 and illustrated in Ex. 18.2 in the text.

Selecting a weak acid or weak base to prepare a buffer of a given pH:

The criterion for selecting a weak acid to prepare a buffer of a given acidic pH is:

$$\text{pH desired} = pK_a \pm 1$$

The criterion for selecting a weak base to prepare a buffer of a given basic pH is:

$$\text{pOH desired} = pK_b \pm 1$$

Calculating the pH of a buffer composed of known concentrations of a weak base and its conjugate acid or the [conjugate acid]/[base] ratio required to obtain a buffer of a given basic pH:

Expression 18w can be used for these kinds of calculations. However, the base form of the Henderson-Hasselbach equation

$$\text{pOH} = pK_b + \log \frac{[\text{conjugate acid}]}{[\text{base}]}$$

can also be used for these kinds of calculations.

Knowing when to do stoichiometric calculations before doing equilibrium calculations:

Many of the calculations encountered in this chapter require solution stoichiometric calculations be done before organizing and conducting equilibrium calculations. This differs from the type of equilibrium calculations conducted thus far and usually presents a challenge to students of general chemistry to know when to do stoichiometric calculations first.

The types of calculations encountered in this chapter can be categorized as:

pH at the equivalence point of a strong acid-weak base reaction*

pH at the equivalence point of a strong base-weak acid reaction*

effect of common ion

pH of buffer of given composition

pH of buffer after addition of a strong acid or strong base*

acid-base composition required to obtain buffer of desired pH

pH for the starting point of a titration curve

pH between the starting point and the equivalence point of a titration curve*

pH at the equivalence point of a strong acid-weak base or strong base-weak acid titration curve*

pH beyond the equivalence point of a titration curve*

and the calculations marked by an asterisk (*) always require stoichiometric calculations be done before equilibrium calculations can be organized and conducted whereas those that are not marked with an asterisk usually do not.

CHAPTER 19

PRINCIPLES OF REACTIVITY: PRECIPITATION REACTIONS

Compounds having a solubility of less than about 0.01 mol of dissolved material per liter of solution are generally considered to be insoluble. The primary focus of this chapter is on the equilibria for the limited dissolving of so-called insoluble salts and on the criteria for forming insoluble compounds by precipitation reactions. (Precipitation reactions were first discussed in Chapter 4, and guidelines for predicting the solubilities of salts were given in Fig. 4.7 in the text.) The chapter concludes with a discussion of simultaneous equilibria and the special role of simultaneous equilibria involving carbon dioxide and carbonates in our environment.

19.1 The Solubility Product Constant, K_{sp}

Most salts that are classified as being insoluble in water dissolve to some extent. *The extent to which an insoluble salt dissolves in water at 25°C can be expressed in terms of the equilibrium constant for the dissolving process*. This constant is called the **solubility product constant** and is designated K_{sp}. The solubility product constant expression for the reaction

$$A_xB_y(s) \rightleftharpoons xA^{y+}(aq) + yB^{x-}(aq) \qquad (19a)$$

is

$$K_{sp} = [A^{y+}]^x[B^{x-}]^y \qquad (19b)$$

The concentration of the solid is omitted from the solubility product constant expression in keeping with the principles developed in Section 16.1. ✶ Try preparing a "flash card" showing the equation for the general reaction of Expression 19a and the corresponding general form of the solubility product constant expression and memorizing it. Also try writing the solubility product constant expressions for various salts until you are comfortable doing them and can do them quickly.

There is an important difference between the *solubility* of a salt and the *solubility product* of that salt. The **solubility** of a salt refers to the maximum quantity of salt that can be dissolved in a given volume, the quantity of salt that is actually dissolved in a given volume of a **saturated solution**, expressed in grams per 100 mL, moles per liter, or other units. On the other hand, the **solubility product** is an equilibrium constant. Nevertheless, *there is a connection between these two quantities; if either is known, the other can be calculated.*

19.2 Determining K_{sp} from Experimental Measurements

Exs. 19.1 and 19.2 in the text illustrate the procedure for calculating solubility product constants from experimentally measured solubilities. Notice how *the stoichiometry of the dissolving reaction is used to calculate the concentrations of the ions when the molar solubility of the compound is given.* The dissolving of $PbCl_2(s)$, for example, yields 1 mole of Pb^{2+} ions and 2 moles of Cl^- ions per mole of $PbCl_2$ dissolved, so the molar concentration of Pb^{2+} in solution is equal to the mol/L of $PbCl_2$ dissolved, and the molar concentration of Cl^- is twice that for Pb^{2+}. In addition, the form of the K_{sp} expression for $PbCl_2$, K_{sp} = $[Pb^{2+}][Cl^-]^2$, still requires that the molar concentration of Cl^- be squared to determine the value of the K_{sp}.

19.3 Estimating Salt Solubility from K_{sp}

The *K_{sp} values* for many insoluble salts have been determined using a variety of experimental techniques and have been compiled in Tables like those in Table 19.1 and in Appendix H in the text. These values *can be used to calculate the solubility of a salt*, as illustrated in Exs. 19.3 and 19.4 in the text. Notice the role of the stoichiometry of the dissolving process in these calculations. *The stoichiometry of the dissolving process tells us the number of ions of each type formed by the dissolution process and therefore the powers to which the concentrations of the ions must be raised and also tells the relative concentrations of the ions involved.* Thus, the dissolving of x mol/L of MgF_2 yields x mol/L of Mg^{2+} and 2x mol/L of F^-, so the expression for calculating the **molar solubility** of $MgF_2(s)$ in water at 25°C, given by x, is $[Mg^{2+}][F^-]^2 = (x)(2x)^2 = K_{sp}$. See "Problem Solving Tips and Ideas" 19.1 in the text for further discussion of these concepts.

The form of the expression that is used to calculate the molar solubility of a salt depends on the stoichiometry of the dissolving process, so we can only use K_{sp} values of salts of like stoichiometry to predict relative solubilities. For example, we can use the decreasing K_{sp} values for AgCl, AgBr and AgI to predict their relative solubilities are S(AgI) < S(AgBr) < S(AgCl), but we cannot just compare the K_{sp} value of AgI to the

K_{sp} value of PbI_2 to predict which of these salts will be more soluble. This is because the molar solubility x of AgI is related to its K_{sp} by $(x)(x) = K_{sp}$ and that of PbI_2 is related to its K_{sp} by $(x)(2x)^2 = K_{sp}$.

19.4   Precipitation of Insoluble Salts

The goal of this section is to develop methods for determining whether an insoluble salt will form a precipitate under a given set of concentrations of its constituent ions.

K_{sp} and the Reaction Quotient Q

The form of the reaction quotient expression is identical to the corresponding equilibrium constant expression, as it was in Section 16.3. The only difference is the concentrations used in calculating the value of Q may or may not be those pertaining to equilibrium. However, we know from the discussion in Section 16.3 that **whenever**

$Q = K_{sp}$, *the system is at equilibrium; the solution is saturated.*

This means no more of the insoluble solid will dissolve.

$Q < K_{sp}$, *the system is not at equilibrium; the solution is unsaturated.*

This means more of the insoluble solid will dissolve until $Q = K_{sp}$. It also means that more cation or anion or both can be added to the solution in the form of a soluble salt or salts, and no precipitation will occur until $Q = K_{sp}$.

$Q > K_{sp}$, *the system is not at equilibrium; the solution is supersaturated.*

This means the concentrations of the ions in solution are too high, and precipitation will occur until $Q = K_{sp}$.

* Try preparing a "flash card" summarizing these three possible Q versus K conditions and their meaning in terms of the solubility of salts and memorizing it.

K_{sp} and Precipitation Reactions

We can use a comparison of Q versus K_{sp} to determine (1) if a precipitate will form when the concentrations of the ions are known or can be calculated and (2) what concentration of an ion is required to begin the precipitation of an insoluble salt when the concentration of the other ion is known.

Exs. 19.6 and 19.8 in the text illustrate the use of initial concentrations of ions to determine whether $Q > K_{sp}$ and thus whether precipitation will occur. **Notice that when solutions are mixed,** as in 19.8, **it is first necessary to use the dilution formula** (Expression 5f) **to calculate the concentrations of the ions in the combined volume before calculating Q.**

Ex. 19.7 in the text illustrates calculating the concentration of an ion necessary to cause Q to equal K_{sp} and cause precipitation to begin. It also involves continuing to add the reactant that contains that ion to the solution to cause precipitation to continue and then stopping when the concentration of the added ion reaches a given value to calculate the concentration of the original ion remaining in solution. ✻ Try practicing the types of problems illustrated in Exs. 19.7 and 19.8 until you are comfortable doing them and can do them quickly.

19.5 Solubility and the Common Ion Effect

The addition of a soluble salt that contains an ion that is part of ("common to") the equilibrium for the dissolving of an insoluble salt lowers the solubility of the insoluble salt. This is another example of the "common ion" effect that was first introduced in 18.2.

Exs. 19.9 and 19.10 in the text illustrate the procedure for calculating the solubility of a salt in the presence of a common ion from another source. **There are two things you should notice about the solution for Ex. 19.9: (1) The concentration of the common ion is included in the ICE table as an initial concentration, as was done in the solution for Ex. 18.3. (2) The concentration of the common ion added is much larger than the concentration of that ion coming from the insoluble salt, so the concentration of ion coming from the insoluble salt can be mathematically neglected with respect to the concentration of the common ion.** This is almost always the case, but you should always check the approximation.

19.6 Solubility, Ion Separations, and Qualitative Analysis

Schemes can be devised to separate ions based on differing solubilities. These can take either of two forms:

- Schemes can take advantage of differing solubilities of two salts having either a common cation and differing anion, as in Ex. 19.11 in the text, or differing cations and a common anion, as with the discussion of AgCl and $PbCl_2$ in the text.

• Schemes can take advantage of having several cations and using different anions to form precipitates to separate and identify the cations, as with the discussion of Ag^+, Pb^{2+}, Cd^{2+} and Cu^{2+} in the text, or of having several anions and using different cations to form precipitates to separate and identify the anions. These kinds of schemes are often used to separate and identify the components of an "unknown" mixture by comparing reactions of the mixture to reactions of a mixture of all the possible components used as knowns in what is called "**qualitative analysis**."

Notice that Ex. 19.11 asks you to determine whether AgCl or Ag_2CrO_4 will precipitate first on addition of $AgNO_3$. This can be done by calculating the $[Ag^+]$ required to <u>begin</u> precipitation of AgCl and Ag_2CrO_4, given the concentrations of Cl^- and CrO_4^{2-} in the solution and the K_{sp} values for both AgCl and Ag_2CrO_4 (Appendix H). The second portion asks you to calculate the $[Cl^-]$ when Ag_2CrO_4 just <u>begins</u> to precipitate. You need to realize that after the AgCl begins to precipitate, $AgNO_3$ is still added to the solution until $[Ag^+]$ reaches the value needed to cause Ag_2CrO_4 to precipitate. Meanwhile, AgCl continues to precipitate decreasing $[Cl^-]$. To calculate the $[Cl^-]$ remaining when Ag_2CrO_4 <u>begins</u> to precipitate, one uses the $[Ag^+]$ required for Ag_2CrO_4 to <u>begin</u> to precipitate with the K_{sp} for AgCl to solve for $[Cl^-]$. ✳ Try preparing a "flash card" outlining the steps to be taken to solve this type of problem and memorizing it. Also try practicing doing this type of problem until you feel comfortable doing them and can do them quickly.

19.7 Simultaneous Equilibria

In many instances two or more equilibrium processes occur at the same time in solution. Chemists characterize these situations as examples of **simultaneous equilibria**.

One example of simultaneous equilibria involves the addition of a reagent to a saturated solution of an insoluble salt converting the salt to a second less soluble salt. The text illustrates the addition of a few drops of K_2CrO_4 to a saturated solution of $PbCl_2$ converting the white $PbCl_2$ to yellow $PbCrO_4$ (margin photo on p. 901 in the text). The reaction that takes place can be considered to the sum of two other reactions, namely

---

Dissolving of PbCl₂:

$$PbCl_2(s) \rightleftharpoons Pb^{2+}(aq) + 2\ Cl^-(aq) \qquad K_1 = K_{sp} = 1.7 \times 10^{-6} \qquad (19c)$$

Precipitation of PbCrO₄:

$$Pb^{2+}(aq) + CrO_4^{2-}(aq) \rightleftharpoons PbCrO_4(s) \qquad K_2 = \frac{1}{K_{sp}} = \frac{1}{1.8 \times 10^{-14}} \qquad (19d)$$

Net reaction:

$$PbCl_2(s) + CrO_4^{2-}(aq) \rightleftharpoons PbCrO_4(s) + 2\ Cl^-(aq) \qquad (19e)$$

$$K_{net} = K_1 \cdot K_2 = 9.4 \times 10^8$$

so **the large value for K_{net} indicates the reaction should be product-favored.**

Other classes of simultaneous equilibria are discussed in the next two sections.

19.8 Solubility and pH

The solubility of metal hydroxides in acid solutions is driven by the union of hydronium ions from the acids and hydroxide ions from the equilibrium of the insoluble hydroxide. For example, the dissolving of magnesium hydroxide in an acid solution can be thought to occur via:

Dissolving of Mg(OH)₂:

$$Mg(OH)_2(s) \rightleftharpoons Mg^{2+}(aq) + 2\ OH^-(aq) \qquad K_{sp} = 1.5 \times 10^{-11} \qquad (19f)$$

Union of hydronium ion and hydroxide ion:

$$H_3O^+(aq) + OH^-(aq) \rightleftharpoons 2\ H_2O(\ell) \qquad K = \frac{1}{K_w} = 1.0 \times 10^{14} \qquad (19g)$$

Net reaction:

$$Mg(OH)_2(s) + 2\ H_3O^+(aq) \rightleftharpoons Mg^{2+}(aq) + 4\ H_2O(\ell) \qquad (19h)$$

$$K_{net} = K_{sp} \cdot K^2 = 1.5 \times 10^{17}$$

where K^2 is used to calculate K_{net}, because it is necessary to multiply 19g by two to obtain the net reaction.

The solubility of metal sulfides and of metal salts with anions

that can act as weak bases (e.g. acetate, carbonate, cyanide, fluoride and phosphate) are enhanced by anion hydrolysis reactions, which can be considered to follow the dissolving process. For example, the solubility of lead sulfide

$$PbS(s) \rightleftharpoons Pb^{2+}(aq) + S^{2-}(aq) \qquad (19i)$$

is increased by

$$S^{2-}(aq) + H_2O(\ell) \rightleftharpoons HS^-(aq) + OH^-(aq) \qquad (19j)$$

having $K_b = 1 \times 10^5$. It is further enhanced by H_3O^+ from strong acids reacting with the OH^- to form the weak electrolyte H_2O.

As a matter of environmental concern, the action of strong acids on limestone ($CaCO_3$) buildings and statutes can be explained by the reaction of H_3O^+ with OH^- formed by hydrolysis of CO_3^{2-}. This system is discussed in detail on p. 902 of the text.

19.9 Solubility and Complex Ions

The solubility of insoluble salts can also be enhanced by the formation of complex ions of the metal ion in Lewis acid-base reactions. For example, AgCl readily dissolves in solutions containing NH_3 due to formation of $Ag(NH_3)_2^+$.

Dissolving of AgCl:

$$AgCl(s) \rightleftharpoons Ag^+(aq) + Cl^-(aq) \qquad K_{sp} = 1.8 \times 10^{-10} \quad (19k)$$

Complexing of Ag^+:

$$Ag^+(aq) + 2\ NH_3(aq) \rightleftharpoons Ag(NH_3)_2^+(aq) \qquad K = 1.6 \times 10^7 \quad (19\ell)$$

Net Reaction:

$$AgCl(s) + 2\ NH_3(aq) \rightleftharpoons Ag(NH_3)_2^+(aq) \qquad K_{net} = K_{sp} \cdot K \quad (19m)$$
$$= 2.9 \times 10^{-3}$$

Indeed, Fig. 19.8 in the text shows a series of reactions involving sequential formation of insoluble silver salts and soluble silver ion complexes. The last reaction is of commercial interest, because it is the reaction that is used to remove unexposed AgBr from black and white photographic film to avoid having the entire film exposed during the developing process.

> Note: Each of the examples of simultaneous equilibria mentioned in Sections 19.7-9 involve the application of LeChatelier's principle. In each case, the removal of a reactant of the first equilibria by the reaction of the second equilibria drives the net reaction towards the product side.

19.10 Equilibria in the Environment: Carbon Dioxide and Carbonates

This section gives some interesting information concerning the roles of CO_2 and $CaCO_3$ in the environment. The reaction of CO_2 with H_2O in the atmosphere causes the formation of H_2CO_3 which ionizes to give H_3O^+ and HCO_3^- and thereby causes the pH of rain in nonpolluted areas to be about 5.6. On the other hand, the hydrolysis of CO_3^{2-} formed by the dissolving of the slightly soluble $CaCO_3$ (limestone) of the oceans gives HCO_3^- and OH^-, which react with acids produced by undersea volcanic activity or other natural processes and thus moderate the pH of the sea. Lastly, it is noted that the amount of CO_2 in the atmosphere has not increased as rapidly as thought, because the increase in the partial pressure of CO_2 above the oceans has caused the solubility of CO_2 in the oceans to increase in keeping with Henry's law (Expression 14h).

19.11 Key Expressions

General dissolving of insoluble salt and solubility product expressions:

$$A_xB_y(s) \rightleftharpoons xA^{y+}(aq) + yB^{x-}(aq)$$

and $$K_{sp} = [A^{y+}]^x[B^{x-}]^y$$

19.12 Hints for Answering Questions and Solving Problems

Using stoichiometric coefficients to express equilibrium concentrations:

Stoichiometric coefficients for dissolving reactions can only be used to express equilibrium concentrations <u>when the ion comes from the dissolving salt</u>. One cannot use these coefficients with concentrations of ions coming from other sources, such as common ion sources. For example, if the common ion source for Ex. 19.10 in the text were 0.10 M $AgNO_3$, the initial concentration to be used for Ag^+ in the ICE table would be 0.10, because the 0.10 M Ag^+ comes from the 0.10 M $AgNO_3$, and the equilibrium concentration would be 0.10 + 2x, where 2x reflects the fact that 2 Ag^+ are produced when one Ag_2CrO_4 dissolves.

CHAPTER 20

PRINCIPLES OF REACTIVITY:
ENTROPY and FREE ENERGY

The goal of this chapter is to consider the factors that determine whether reactions are reactant or product-favored and to develop a model that will allow us to predict the same. We will see that there are two tendencies which are operative, namely the tendency for the dispersal of energy and the tendency for the dispersal of matter.

In some cases, these tendencies reinforce one another and in other cases they oppose one another and the outcome is affected by the temperature at which the reaction occurs. So, some reactions are reactant-favored at all temperatures, some are product-favored at all temperatures, some are reactant-favored at low temperatures and product-favored at high temperatures and still others are product-favored at low temperatures and reactant-favored at high temperatures. Thus, you can see why it is important for us to understand the factors affecting the outcomes of chemical reactions. In addition, we would like to be able to express the outcomes of reactions in a quantitative manner, and you will see that we are able to use the principles of thermodynamics to calculate values of equilibrium constants to do so.

20.1 Spontaneous Reactions and Speed: Thermodynamics versus Kinetics

Thermodynamics is the science of energy transfer that enables us to predict whether a reaction can occur to give predominantly products with sufficient time. However, thermodynamics cannot tell us anything about the amount of time that will be required. The study of rates of reactions, and why some are slow and others are fast, is called **kinetics**, the topic of Chapter 15.

In Chapter 6 we defined a reaction that can occur to give predominantly products with sufficient time as a **product-favored reaction.** Conversely, a reaction that can only yield mostly reactants with time is said to be a **reactant-favored reaction**. However, it should be noted that chemists often use the term **spontaneous** to describe product-favored reactions and the term **nonspontaneous** to describe reactant-favored reactions. Unfortunately, the term "spontaneous" usually implies the process of interest occurs quickly, thus mixing the concepts of thermodynamics and kinetics. This is why the text uses the terms

product-favored and **reactant-favored** to clearly separate the concept of reaction probability from the concept of reaction speed.

The "Closer Look" on p. 916 in the text contains a summary of other terms that were used in Chapter 6 and will be used again in this chapter. * Try preparing a separate "flash card" for each term placing the term on one side and the meaning on the other side and memorizing them.

20.2 Directionality of Reactions: Entropy

On page 289 in the text, it was stated that *most exothermic reactions are product-favored at room temperature. During an exothermic reaction, the chemical potential energy that is stored in a relatively few atoms and molecules of the reactants is dispersed over the much larger number of atoms and molecules of the products and surroundings. Energy dispersal is always favored, because it is much more probable that energy will be dispersed than it is that energy will be concentrated.*

There is also a tendency for highly concentrated matter to disperse, as illustrated in Fig. 20.3 in the text. Thus, *there are two ways that the final state of a system can be more probable than its initial state: (1) having energy dispersed over a greater number of molecules and (2) having the atoms and molecules themselves become more disordered. If the reaction is*

• exothermic and leads to more disorder, there is dispersal of both energy and matter, and the reaction is product-favored at all temperatures.

• exothermic but leads to more order, the reaction is likely to be product-favored at room temperature and reactant-favored at high temperatures.

> *This means the dispersal of energy is usually more important than the dispersal of matter at room temperature, but the opposite is true at higher temperatures.*

• endothermic but leads more disorder, the reaction is likely to be reactant-favored at room temperature and product-favored at higher temperatures.

• endothermic and leads to more order, there is neither dispersal of energy or matter, and the reaction is reactant-favored at all temperatures.

* Try preparing a separate "flash card" for each of these four possibilities and memorizing them.

Entropy: A Measure of Matter Dispersal or Disorder

The dispersal or disorder in a sample of matter can be measured using a calorimeter (Fig. 6.17 in the text), the same instrument that can be used to measure the enthalpy change that occurs when a reaction takes place. The result is a thermodynamic state function that is called **entropy** and symbolized by S. *The measurement of entropy is based on the fact that all translational motion ceases at the absolute zero of temperature, 0 K or -273.15°C, and there is no disorder in a pure, perfectly formed crystalline substance at this temperature.* This is why the **Third Law of Thermodynamics** states the entropy of a pure, perfectly formed crystalline substance is zero at the absolute zero of temperature: S (0 K) = 0.

When energy is added to a substance in very small increments as heat, so that the temperature change is small, the entropy change can be calculated as $\Delta S = q/T$, where T is the absolute temperature at which the heat is added. Starting as close as possible to absolute zero and adding small quantities of energy makes it possible to determine the entropy change for each small increase in temperature. These changes can then be added together to give the total or **absolute entropy** of a substance at any desired temperature. However, the results are usually tabulated as **standard molar entropy** values, corresponding to 1 mol at 298 K and the standard pressure of 1 bar (0.98692 atm), and are expressed in units of joules per kelvin per mole ($J/K\cdot$mol).

Some interesting patterns are seen in entropy data like that given in Table 20.1 in the text.

• *When comparing the same or similar substances, entropy increases from the solid to the liquid to the gaseous state* paralleling the increase in the freedom of the particles to move around and the increase in disorder.

$$S(s) < S(\ell) < S(g)$$

• *When comparing the members of a series of closely related compounds, entropy increases as the molecules become more complex,* because there are more ways for the atoms to be arranged in three-dimensional space with the more complicated molecules. Hence,

$$S(CH_4) < S(CH_3CH_3) < S(CH_3CH_2CH_3)$$

• **When comparing ionic solids, entropy increases as the charges on the ions decrease** and make it easier for the ions to vibrate about their lattice positions. Hence,

$$S(MgO) < S(NaF)$$

• **When a pure solid or liquid dissolves in a solvent, entropy usually increases**, because matter usually becomes more dispersed or scattered as a substance dissolve.

• **When a dissolved gas escapes from a solution, entropy increases**, because the large increase in entropy that accompanies going from the liquid to the gas phase offsets the loss of disorder accompanying the break-up of the mixture of the solution.

These generalizations can in turn be used to predict whether there will be an increase in entropy accompanying physical and chemical changes, as shown on pp. 921 and 922 in the text. **Notice that there is an increase in disorder and increase in entropy whenever there is an increase in the number of moles of gas on moving from reactants to products.** ∗ Try preparing a separate "flash card" for each of these generalizations summarizing the trend in entropies and the reason for the trend and memorizing them.

The signs of ΔS for physical changes of state are particularly easy to predict qualitatively, without calculations, because $S(s) < S(\ell) < S(g)$ for any given pure substance. However, physical changes of state also occur at constant temperatures for pure substances, and enthalpy changes can be measured and tabulated for such changes making it possible to actually calculate the corresponding ΔS values using $\Delta S = q/T$. See pp. 921 and 922 in the text for examples pertaining to changes of state involving water.

Entropy and the Second Law of Thermodynamics

The **Second Law of Thermodynamics** states the total entropy of the universe is continually increasing. **This means that a product-favored reaction is accompanied by an increase in the entropy of the universe; i.e. $\Delta S_{universe}$ is positive.** Thus, **evaluating whether the entropy of the universe will increase during a chemical reaction allows us to predict whether or not reactants will form predominantly products.** Such evaluations are usually done for **standard conditions** (temperature usually = 298 K, pressure = 1 bar and concentrations = 1 molal, which for dilute solutions is virtually 1 molar; see pp. 288 and 916 in the text.), and corrections are usually made for differences from standard conditions.

Calculating whether a reaction will have a positive value for $\Delta S_{universe}$ and therefore be product-favored involves two steps: (1) calculating how much entropy is created by the dispersal of matter, and (2) calculating how much entropy is created by the dispersal of energy. *Both of these calculations are done assuming that reactants at standard conditions are converted to products at standard conditions.*

Calculating the Entropy Change for a System

To calculate the entropy change due to the dispersal of matter during the course of a reaction, ΔS_{system}, we assume that each reactant and each product is present in the mole amount required by its stoichiometric coefficient. We also assume that all substances are present at the temperature specified and at the standard pressure, so the standard entropy values from Table 20.1 and Appendix K will apply. Then we add up all the entropies of the products and subtract all the entropies of the reactants to see whether there is an increase or decrease in the entropy of the system. That is, we use

$$\Delta S^{\circ}_{system} = \Sigma \ S^{\circ}(products) - \Sigma \ S^{\circ}(reactants) \qquad (20a, 20.1)$$

where the capital Greek letter sigma, Σ, means to "take the sum."

Applying Expression 20a to the general equation aA + bB $\rightarrow$ cC + dD gives:

$$\Delta S^{\circ}_{rxn} = cS^{\circ}[C] + dS^{\circ}[D] - aS^{\circ}[A] - bS^{\circ}[B] \qquad (20b)$$

as illustrated in Ex. 20.1 in the text. *Notice that it is necessary to multiply the absolute entropy value of each substance by the numbers of moles of that substance in the balanced equation, because S^{0} values are tabulated in terms of J/K·mol (See Table 20.1 and Appendix K in the text.). The entropy change that is calculated for the system using this expression, ΔS°_{rxn}, is the entropy change that occurs when reactants are completely converted to products under standard conditions.*

Calculating the Entropy Change in the Surroundings

The entropy change created by the dispersal of energy by a chemical reaction, the entropy change for the surroundings, can be calculated by calculating ΔH° for the reaction (using data from Tables like 6.2 or Appendix K with Expression 6h) and by assuming that this quantity of energy is transferred to or from the surroundings. If the energy transfer is slow and occurs at a constant temperature, the entropy change for the surroundings can be calculated using

---

$$\Delta S^{\circ}_{\text{surroundings}} = \frac{q_{\text{surroundings}}}{T} = \frac{-\Delta H^{\circ}_{\text{system}}}{T} \quad (20c)$$

where $-\Delta H^{\circ}_{\text{system}}$ is the energy transferred <u>out of the system</u> and into the surroundings. This means that for an exothermic reaction ($\Delta H^{\circ}_{\text{system}}$ = a negative value), energy will be transferred into the surroundings and there will be an increase in the entropy of the surroundings, as we should expect.

Calculating the Total Entropy Change for System and Surroundings

The total entropy change for a process conducted under standard conditions, $\Delta S^{\circ}_{\text{universe}}$, is given by the sum of the standard entropy change for the system, $\Delta S^{\circ}_{\text{system}}$, and the standard entropy change for the surroundings, $\Delta S^{\circ}_{\text{surroundings}}$.

$$\Delta S^{\circ}_{\text{universe}} = \Delta S^{\circ}_{\text{system}} + \Delta S^{\circ}_{\text{surroundings}} \quad (20d)$$

When $\Delta S^{\circ}_{\text{universe}}$ is positive, there is an increase in the entropy of the universe, and the process is product-favored in accord with the Second Law of Thermodynamics.

The sign of $\Delta S^{\circ}_{\text{universe}}$ can also be predicted qualitatively, without calculations, using the information that is summarized in Table 20.2 in the text, which is just a more formal summary of the one we gave on p. 272 and includes the signs of $\Delta H^{\circ}_{\text{system}}$ and $\Delta S^{\circ}_{\text{system}}$. The entries in this table are based on the fact that Expression 20c can be rewritten as

$$\Delta S^{\circ}_{\text{universe}} = \Delta S^{\circ}_{\text{system}} - \frac{\Delta H^{\circ}_{\text{system}}}{T} \quad (20e)$$

and the sign of $\Delta S^{\circ}_{\text{universe}}$ predicted by substituting the signs of $\Delta S^{\circ}_{\text{system}}$ and $\Delta H^{\circ}_{\text{system}}$ into Expression 20d. Notice that the form of Expression 20e tells us that <u>when the signs of $\Delta S^{\circ}_{\text{system}}$ and $\Delta H^{\circ}_{\text{system}}$ are either both negative or both positive, the outcome depends on the temperature</u>. Also notice that the form of Expression 20e tells that the higher the temperature, the less important the $\Delta H^{\circ}_{\text{system}}$ term is. This is why we often say reactions are enthalpy driven at low temperatures, like room temperature, and entropy driven at high temperatures. Of course, this assumes that $\Delta S^{\circ}_{\text{system}}$ and $\Delta H^{\circ}_{\text{system}}$ do not change much with temperature, and this is usually a good assumption. ✻ Try preparing a "flash card" summarizing the information of Table 20.2 in the text and memorizing it; under the product-favored column just use "At lower temperatures." for the entry for when $\Delta H^{\circ}_{\text{system}}$ and $\Delta S^{\circ}_{\text{system}}$ are both negative and "At higher temperatures." for the entry for when $\Delta H^{\circ}_{\text{system}}$ and $\Delta S^{\circ}_{\text{system}}$ are both positive.

20.3 Gibbs Free Energy

J. Willard Gibbs (1839-1903) of Yale University introduced a new thermodynamic state function for the system, now called the **Gibbs free energy** and given the symbol **G**, that can be used to determine whether a process will be reactant or product-favored. *Gibbs defined his free energy function as*

$$\Delta G_{system} = \Delta H_{system} - T\Delta S_{system} \tag{20f}$$

which under standard conditions becomes

$$\Delta G^{\circ}_{system} = \Delta H^{\circ}_{system} - T\Delta S^{\circ}_{system} \tag{20g}$$

Notice that rearranging Expression 20e gives

$$-T\Delta S^{\circ}_{universe} = \Delta H^{\circ}_{system} - T\Delta S^{\circ}_{system} \tag{20h}$$

so by comparison with Expression 20f

$$\Delta G^{\circ}_{system} = -T\Delta S^{\circ}_{universe} \tag{20i}$$

This means $\Delta G^{\circ}_{system}$ must be negative for $\Delta S^{\circ}_{universe}$ to be positive and the process to be product-favored. It furthermore means that Expression 20g can be used to reach the same conclusions with regard to whether processes are reactant or product-favored processes as was presented in Table 20.2 in the text. Indeed, Expression 20g is the expression that is usually used to rationalize the temperature dependence of the "spontaneity" of reactions.

Calculating the Free Energy Change for a Reaction, ΔG°_{rxn}

Enthalpy and entropy changes can be calculated for chemical reactions using ΔH°_f and S° values from Tables 6.2 and 20.1 and Appendix K with Expressions 6h (6.1) and 20a (20.1). The resulting values can then be used with Expression 20g (20.2) to calculate ΔG°_{rxn} (= $\Delta G^{\circ}_{system}$), as in Ex. 20.2 in the text. Also notice how the favorable enthalpy change for the reaction in Ex. 20.2 in the text dominates the unfavorable entropy change at the low temperature of 25°C, so the reaction really is enthalpy driven at 25°C.

Standard Free Energy of Formation, ΔG°_f

The standard free energy change for the formation of one mole of a compound from its elements, with the reactants and the product in their standard states, is called the **standard molar free**

energy of formation for the compound. *Elements in their standard states have ΔG_f^o values of 0, for the same reason that they have ΔH_f^o values of 0;* see p. 75 and also the listing of the standard states of the elements on pp. 79 and 80.

The equation which relates ΔG_{rxn}^o to ΔG_f^o values is analogous to the equation which relates ΔH_{rxn}^o to ΔH_f^o values (6h, 6.6). It is

$$\Delta G_{rxn}^o = \Sigma \ \Delta G_f^o(\text{products}) \ - \ \Sigma \ \Delta G_f^o(\text{reactants}) \qquad (20j, 20.3)$$

*and applying this expression to the general equation aA + bB →
cC + dD gives:*

$$\Delta G_{rxn}^o = c\Delta G_f^o[\text{C}] + d\Delta G_f^o[\text{D}] - a\Delta G_f^o[\text{A}] - b\Delta G_f^o[\text{B}] \qquad (20k)$$

*Notice once again how it is necessary to multiply the ΔG_f^o value
for each substance by the numbers of moles of that substance in
the balanced equation, because ΔG_f^o values correspond to forming
just one mole of compound and are, therefore, tabulated in terms
of kJ/mol (See Table 20.3 and Appendix K in the text.).* The
application of Expression 20j is illustrated in Ex. 20.3 in the
text, and the meaning of the term "free" energy is discussed in
Problem Solving Tips and Ideas 20.1 in the text. **Free energy**
refers to the interplay between the change in thermal energy and
energy required to bring order to the system and thus represents
the maximum net amount of useful work that can be obtained from
a reaction. ✳ Try preparing a "flash card" giving this statement about free energy
and memorizing it.

Product-Favored or Reactant Favored

*Expression 20g tells us the sign for ΔG_{rxn}^o ultimately depends on
the signs for ΔH_{rxn}^o and ΔS_{rxn}^o and the temperature T at which the
reaction is conducted. If ΔG_{rxn}^o is positive at the temperature of
interest, the process is predicted to be reactant-favored, and
if it is negative, it is predicted to be product-favored.* The
text illustrates calculations for processes involving the
various combinations of signs for ΔH_{rxn}^o and ΔS_{rxn}^o. The dissolving
of ammonium nitrate, NH_4NO_3, is of interest, because it is an
example of a process that is entropy driven at the low temperature of 25°C.

Free Energy and Temperature

The qualitative influence of temperature on determining whether
a process is reactant or product-favored has been previously
discussed on pp. 272, 276 and 277. However, *here we wish to
outline a procedure for determining the temperature range of
"spontaneity," the temperature range over which reactions are
either reactant or product-favored.*

• First determine the sign of ΔG°_{rxn} at 298 K (25°C) to determine whether the reaction is reactant or product-favored.

• Next determine the temperature at which there is no net driving force in either direction under standard conditions by setting $\Delta G^{\circ}_{rxn} = 0 = \Delta H^{\circ}_{rxn} - T\Delta S^{\circ}_{rxn}$ and solving for T.

• Finally, examine the trend in the sign of ΔG°_{rxn} from 298 K to the temperature T calculated in the step above.

    • If ΔG°_{rxn} goes from negative to zero as the temperature goes from 298 K to T, it will become positive at temperatures $>T$, and the reaction will change from being product-favored at temperatures $<T$ to being reactant-favored at temperatures $>T$.

    • If ΔG°_{rxn} goes from positive to zero as the temperature goes from 298 K to T, it will become negative at temperatures $>T$, and the reaction will change from being reactant-favored at temperatures $<T$ to being product-favored at temperatures $>T$.

✻ Try preparing a "flash card" summarizing the procedure for predicting the temperature range of "spontaneity" for chemical reactions and memorizing it. Also try doing this type of calculation until you are comfortable doing them and can do them quickly.

20.4   Thermodynamics and the Equilibrium Constant

A product-favored reaction proceeds largely to products, but some reactants can remain at equilibrium. A reactant-favored reaction proceeds only partially to products before achieving equilibrium. The question is thus: How is ΔG°_{rxn} related to the conditions at equilibrium?

The free energy change pertaining to any point in a reaction, ΔG, is related to the free energy change pertaining to standard conditions, ΔG°, by

$$\Delta G = \Delta G^{\circ} + RT\ln Q \qquad (20\ell)$$

where R is the universal gas constant expressed as 8.3145 $J/K\cdot mol$, T is the temperature in kelvins and Q is the thermodynamic reaction quotient for the reaction (16.3).

At equilibrium Δ*G* = 0 and *Q* = *K*, *so substituting into Expression 20ℓ gives*

$$0 = \Delta G^{\circ} + RT\ln K \tag{20m}$$

and rearranging Expression 20m yields

$$\Delta G^{\circ} = -RT\ln K \tag{20n, 20.4}$$

where K is considered to be the **thermodynamic equilibrium constant** for the reaction. *For reactions involving only gases, the thermodynamic equilibrium constant is K_p. For compounds and ions in solution, the thermodynamic equilibrium constant is K_c.*

The qualitative relation between free energy and equilibrium is illustrated graphically in Fig. 20.9 in the text. The left side of each graph gives the total free energy for the reactants, the right side gives the total free energy for the products and the difference between the two is ΔG°, which, like K, depends only on temperature.

The diagram on the left in Fig. 20.9 corresponds to a negative value for ΔG°, because $\Sigma\ G^{\circ}$(products) < $\Sigma\ G^{\circ}$(reactants) and ΔG° = $\Sigma\ G^{\circ}$(products) - $\Sigma\ G^{\circ}$(reactants), *and a K that is greater than one. Conversely, the diagram on the right corresponds to a positive value for ΔG° and a K that is less than one*, where the relation between ΔG° and K can be determined from the mathematical form of Expression 20n. On the other hand, *we should expect the equilibrium point to be towards the product side when ΔG° is negative and toward the reactant side when ΔG° is positive, since these are the signs associated with product and reactant-favored reactions, respectively.* ✶ Try preparing separate "flash cards" showing labelled Figs. 20.9a and 20.9b and memorizing them.

Lastly, notice the comparisons of Q versus K given on the graphs of Fig. 20.9. *You should recall from Chapter 16 that*

   • when $Q < K$, the reaction proceeds from the reactant side to the product side to get to the equilibrium point, because the ratio of product concentrations to reactant concentrations, each raised to the power of their stoichiometric coefficient, is too low.

   • when $Q = K$, the system is at equilibrium.

   • when $Q > K$, the reaction proceeds from the product side to the reactant side to get to the equilibrium point, because the ratio of product concentrations to reactant concentrations, each raised to the power of their stoichiometric coefficient, is too high.

The quantitative relation between ΔG^o_{rxn} and the point of chemical equilibrium, as expressed by the value of K, is given by Expression 20n. Applications of Expression 20n involve: (1) the calculation of equilibrium constants from thermodynamic data and (2) the calculation of ΔG^o_{rxn} from an experimental value of K. These applications are illustrated in Exs. 20.4 and 20.5 in the text, though you should realize that there are cases where it is necessary to use either Expression 20g or 20j to calculate ΔG^o_{rxn} to be able to calculate the value of K.

Lastly, **Ex. 20.6 in the text offers an excellent summary example of the use of concepts and calculations encountered in this chapter.** So, you should pay particular attention to this Example.

20.6 Key Expressions

Entropy of change of state - enthalply of change of state expression:

$$\Delta S = q/T$$

Entropy of reaction - absolute entropy expression:

$$\Delta S^o_{system} = \Sigma \ S^o(products) - \Sigma \ S^o(reactants)$$

Entropy change of universe expressions:

$$\Delta S^o_{universe} = \Delta S^o_{system} + \Delta S^o_{surroundings}$$

$$\Delta S^o_{universe} = \Delta S^o_{system} - \frac{\Delta H^o_{system}}{T}$$

Free energy - enthalpy - entropy of reaction expression:

$$\Delta G^o_{system} = \Delta H^o_{system} - T\Delta S^o_{system}$$

Free energy of reaction - free energy of formation expression:

$$\Delta G^o_{rxn} = \Sigma \ \Delta G^o_f(products) - \Sigma \ \Delta G^o_f(reactants)$$

Free energy of reaction - thermodynamic equilibrium constant expressions:

$$\Delta G^o = -RT\ln K$$

$$\ln K = -\Delta G^o/RT \quad and \quad K = antiln(-\Delta G^o/RT) = e^{-\Delta G^o/RT}$$

20.7 Hints for Answering Questions and Solving Problems

Recalling the three laws of thermodynamics:

First law: The total energy of the universe is constant.

Second law: The total entropy of the universe is always increasing.

Third law: The entropy of a pure, perfectly formed crystalline substance is zero at the absolute zero of temperature.

✳ Try preparing a "flash card" summarizing these laws and memorizing it.

Working with Expressions 20g and 20n:

Caution must used when working with Expressions 20g and 20n, because these expressions contain quantities which are expressed in differing units. In Expression 20g, ΔG°_{system} and ΔH°_{system} are usually expressed in kJ whereas ΔS°_{system} is usually expressed in J/K. Hence, it is desirable to convert ΔS°_{system} to kJ/K for use in Expression 20g. Similarly, ΔG° and R appear in Expression 20n, and ΔG° is usually expressed in kJ and R in J, so it is desirable to convert R to 8.3145×10^{-3} kJ/K·mol for use in Expression 20n.

Recalling the relation between ΔG° and K based on Expression 20n:

For the reaction $aA + bB \rightleftharpoons cC + dD$ with $Q = [C]^c[D]^d/[A]^a[B]^b$,

ΔG°	K	Product Formation
<0	>1	Products favored over reactants at equilibrium
=0	=1	At equilibrium with $[C]^c[D]^d=[A]^a[B]^b$; very rare
>1	<1	Reactants favored over products at equilibrium

✳ Try preparing a "flash card" summarizing these relations and memorizing it.

Understanding the relation between ΔG° and ΔG:

The value of ΔG° for a reaction is fixed at a given temperature and determines the position of equilibrium for the reaction at that temperature, because $\Delta G^\circ = -RT\ln K$. The value of ΔG, however, changes during the course of reaction, because Q changes. This determines whether the system is at equilibrium at any given point in the reaction. Furthermore, if the system is not at equilibrium, the value of ΔG tells us which direction the reaction will need to go to get to equilibrium, because all reactions proceed in the direction that gives $\Delta G = 0$ at equilibrium. We can, for example, substitute $-RT\ln K$ for ΔG° in Expression 20ℓ to obtain $\Delta G = RT\ln(Q/K)$, which tells us ΔG is negative when $Q < K$ and the reaction will proceed in the forward direction to get to equilibrium. When $Q > K$, ΔG is positive for the forward direction but negative for the reverse direction, so the reaction proceeds in the reverse direction to get to equilibrium. The slopes of the free energy plots shown in Fig. 20.9 also give the sign of ΔG and tell us this same information.

CHAPTER 21

PRINCIPLES OF REACTIVITY:
ELECTRON TRANSFER REACTIONS

Electron transfer reactions, called oxidation-reduction reactions, are involved in the conversion of the oxygen we breathe to the carbon dioxide and water produced during the combustion of food within us and the conversion of water to oxygen in green plants by photosynthesis. Oxidation-reduction reactions are also involved in the production of steel from iron ore, the corrosion of steel and the prevention of corrosion. In addition, oxidation-reduction reactions are used by batteries to generate electrical energy and, conversely, electrical energy from power plants can be used to produce the active metals of Groups 1A and 2A and aluminum and to purify metals, such as copper for use in electrical wiring, by electron transfer reactions. Thus, oxidation-reduction reactions are an essential part of life.

We will begin our consideration of these important reactions by describing some special techniques for balancing equations for oxidation-reduction reactions.

21.1 Oxidation-Reduction Reactions

Oxidation-reduction reactions, also called **redox reactions**, occur by electron transfer. Perhaps you recall the introduction to this major class of reactions in Section 4.10. You may wish to review that material, especially the definitions of the terms **oxidation, reduction, reducing agent, oxidizing agent and oxidation number** and the "Guidelines for Determining Oxidation Numbers" given on p. 191 in the text. ✷ Try preparing "flash cards" for these terms, if you haven't already done so, and memorizing them.

Balancing Equations for Oxidation-Reduction Reactions

The equations for some redox reactions can be balanced by the inspection or "trial and error" methods. However, the equations for most redox reactions require the use of a more systematic method. The goal of this Section is to develop such a method.

Consider the relatively simple reaction of zinc metal with copper(II) ions that is illustrated in Fig. 21.2 in the text. *The zinc metal reacts with the copper(II) ions to give zinc ions in solution and copper metal. During the course of this reaction,*

• Zn loses electrons, is oxidized to Zn^{2+} (See Fig. 3.10 in the text.) and Cu^{2+} is the oxidizing agent.

• Cu^{2+} gains electrons, is reduced to Cu and Zn is the reducing agent.

The equation for the overall reaction can be written as the sum of the equations for the two **half-reactions**: one for the oxidation of the zinc and the other for the reduction of the copper(II) ions.

Oxidation: $\qquad\qquad\qquad Zn(s) \rightarrow Zn^{2+}(aq) + 2\ e^-$
Reduction: $\qquad Cu^{2+}(aq) + 2\ e^- \rightarrow Cu(s)$

Net reaction: $\quad Zn(s) + Cu^{2+}(aq) \rightarrow Zn^{2+}(aq) + Cu(s)$

The half-reactions and net equation each have an atom balance and a charge balance and are therefore balanced equations.

The procedure that was used to obtain the balanced equation for the reaction of zinc with copper(II) ions illustrates a portion of the more general procedure for balancing equations for redox reactions. This involves

1) identifying the reaction as an oxidation-reduction reaction based on changes in oxidation numbers,

2) identifying the element(s) oxidized on the basis of an increase in oxidation number and the substance that causes the oxidation, the oxidizing agent, and the element(s) reduced on the basis of a decrease (reduction) in oxidation number and the substance that causes the reduction, the reducing agent,

3) separating the reaction into half-reactions by examining the forms of each element on the left and right hand sides of the starting equation, the **skeleton equation**,

4) labelling the half-reactions as the oxidation and reduction half-reactions based on the increase and decrease, respectively, in the oxidation numbers of an element appearing in the half-reactions,

5) balancing each half-reaction for mass by achieving an atom balance,

6) balancing each half-reaction for charge by adding electrons to the side of the half-reaction that has the more positive overall charge,

---

7) multiplying each half-reaction by an appropriate factor to ensure that the number of electrons gained by the oxidizing agent will equal the number of electrons lost by reducing agent,

8) adding the multiplied half-reactions together to produce the balanced overall equation and reducing it to its simplest form by canceling any species appearing on both sides of the equation, as you would reduce $6x + y = 3x + 3y$ to $3x = 2y$, and

9) checking the overall equation to ensure that it has both an atom balance and charge balance.

Let's apply this procedure to balancing the equation for the aqueous solution reaction of iodide ions with iron(III) ions to produce iodine and iron(II) ions:

1,2) The starting equation, the skeleton equation, for this reaction is

$$I^-(aq) + Fe^{3+}(aq) \rightarrow I_2(aq) + Fe^{2+}(aq)$$

Oxidation nos. -1 $+3$ 0 $+2$

and a consideration of the oxidation numbers placed under the elements indicates there is an increase in oxidation number for the iodine and a decrease in oxidation number for the iron, so the reaction is indeed a redox reaction. In fact,

• I^- loses electrons, is oxidized to I_2 and Fe^{3+} is the oxidizing agent.

• Fe^{3+} gains electrons, is reduced to Fe^{2+} and I^- is the reducing agent.

3,4) The half-reactions necessarily involve the iron species in one half-reaction and the iodine species in the other. Based on the identification of the oxidation of I^- and the reduction of Fe^{3+}, they are:

Oxidation: $I^-(aq) \rightarrow I_2(aq)$
Reduction: $Fe^{3+}(aq) \rightarrow Fe^{2+}(aq)$

5) To achieve an atom balance in each equation, it is necessary to multiply I^- by 2 giving

Oxidation: $2\ I^-(aq) \rightarrow I_2(aq)$
Reduction: $Fe^{3+}(aq) \rightarrow Fe^{2+}(aq)$

6) To achieve a charge balance in each equation, it is necessary to add two electrons to the right side of the oxidation half-reaction to make the total charge on each side -2 and one electron to the left side of the reduction half-reaction to make the total charge on each side +2.

Oxidation: $2 \, I^-(aq) \rightarrow I_2(aq) + 2 \, e^-$
Reduction: $Fe^{3+}(aq) + e^- \rightarrow Fe^{2+}(aq)$

> **Note:** Electrons are always added to the right side of an oxidation half-reaction, because oxidation involves the loss of electrons. Similarly, electrons are always added to the left side of a reduction half-reaction, because reduction involves the gain of electrons.

7) To ensure that the number of electrons gained by the oxidizing agent will equal the number of electrons lost by the reducing agent, it is necessary to multiply the reduction half-reaction by 2 giving

Oxidation: $2 \, I^-(aq) \rightarrow I_2(aq) + 2 \, e^-$
Reduction: $2 \, Fe^{3+}(aq) + 2 \, e^- \rightarrow 2 \, Fe^{2+}(aq)$

8) To produce the balanced overall equation, add the multiplied half-reactions together giving

Oxidation: $2 \, I^-(aq) \rightarrow I_2(aq) + 2 \, e^-$
Reduction: $2 \, Fe^{3+}(aq) + 2 \, e^- \rightarrow 2 \, Fe^{2+}(aq)$
Net reaction: $2 \, I^-(aq) + 2 \, Fe^{3+}(aq) \rightarrow I_2(aq) + 2 \, Fe^{2+}(aq)$

9) There are 2 I and 2 Fe on each side of the equation and the total overall charge on each side is +4, so the net equation is balanced.

Ex. 21.1 in the text illustrates another example of balancing an equation for a redox reaction involving just metals and their ions. However, at this point, you are probably wondering why we've placed so much emphasis on developing a systematic method for balancing redox equations, because we've only applied it to two equations that could have been balanced quickly by inspection. Now that we've developed the method, we are ready to move on to balancing more complex redox equations for reactions occurring in acidic or basic solution.

Balancing Equations for Oxidation-Reduction Reactions Occurring in Acidic Solutions:

In acidic solutions, H_3O^+, or simply H^+, and H_2O are able to participate in the reaction in addition to those substances given in the skeleton equation. Thus, to achieve an atom balance for a half-reaction occurring in acidic solution, we

> **5a)** balance all atoms other than oxygen and hydrogen by adjusting the coefficients of the substances in the half-reaction,
>
> **5b)** balance O atoms by adding the appropriate number of H_2O molecules to the side that is deficient in O atoms, and
>
> **5c)** balance H atoms by adding the appropriate number of H^+ ions to the side that is deficient in H atoms.

For example, given the skeleton equation

$$Fe^{2+} + Cr_2O_7^{2-} \rightarrow Fe^{3+} + Cr^{3+} \qquad \text{(acidic solution)}$$

1,2) we should write

$$Fe^{2+} + Cr_2O_7^{2-} \rightarrow Fe^{3+} + Cr^{3+}$$

Oxidation nos. +2   +6,−2   +3   +3

and note that there is an increase in the oxidation number of the iron and a decrease in the oxidation number of the chromium, so the reaction is indeed a redox reaction. In fact,

> • Fe^{2+} loses electrons, is oxidized to Fe^{3+} and $Cr_2O_7^{2-}$ is the oxidizing agent.
>
> • $Cr(+6)$ gains electrons, is reduced to Cr^{3+} and Fe^{2+} is the reducing agent.

> **Note:** The Cr does not exist as an ion having a 6+ charge in $Cr_2O_7^{2-}$, because $Cr_2O_7^{2-}$ is a covalently bonded ion. This is why we've stated $Cr(+6)$, not Cr^{6+}, gains electrons. Also notice that it is the Cr within the $Cr_2O_7^{2-}$ that gains electrons, not the entire $Cr_2O_7^{2-}$. However, we commonly use the entire formula $Cr_2O_7^{2-}$ to represent the formula of the oxidizing agent, because the $Cr(+6)$ does not exist by itself.

3,4) The half-reactions necessarily involve the iron species in one half-reaction and the chromium species in the other. Based on the identification of the oxidation of Fe^{2+} and the reduction of Cr(+6), they are:

Oxidation: $Fe^{2+} \rightarrow Fe^{3+}$
Reduction: $Cr_2O_7{}^{2-} \rightarrow Cr^{3+}$

5a) To achieve an atom balance for atoms other than O and H in each equation, it is necessary to multiply the Cr^{3+} by 2 giving

Oxidation: $Fe^{2+} \rightarrow Fe^{3+}$
Reduction: $Cr_2O_7{}^{2-} \rightarrow 2\ Cr^{3+}$

5b) To achieve an O atom balance in the reduction half-reaction, it is necessary to add 7 H_2O to the right side giving

Oxidation: $Fe^{2+} \rightarrow Fe^{3+}$
Reduction: $Cr_2O_7{}^{2-} \rightarrow 2\ Cr^{3+} + 7\ H_2O$

5c) To achieve a H atom balance in the reduction half-reaction, it is necessary to add 14 H^+ to the left side giving

Oxidation: $Fe^{2+} \rightarrow Fe^{3+}$
Reduction: $Cr_2O_7{}^{2-} + 14\ H^+ \rightarrow 2\ Cr^{3+} + 7\ H_2O$

6) To achieve a charge balance in each equation, it is necessary to add 1 electron the right side of the oxidation half-reaction to make the total charge on each side +2 and 6 electrons to the left side of the reduction half-reaction to make the total charge on each side +6.

Oxidation: $Fe^{2+} \rightarrow Fe^{3+} + e^-$
Reduction: $Cr_2O_7{}^{2-} + 14\ H^+ + 6\ e^- \rightarrow 2\ Cr^{3+} + 7\ H_2O$

7) To ensure that the number of electrons gained by the oxidizing agent will equal the number of electrons lost by the reducing agent, it is necessary to multiply the oxidation half-reaction by 6 giving

Oxidation: $6\ Fe^{2+} \rightarrow 6\ Fe^{3+} + 6\ e^-$
Reduction: $Cr_2O_7{}^{2-} + 14\ H^+ + 6\ e^- \rightarrow 2\ Cr^{3+} + 7\ H_2O$

8) To produce the balanced overall equation, add the multiplied half-reactions together giving

Oxidation: $6\ Fe^{2+} \rightarrow 6\ Fe^{3+} + 6\ e^-$
Reduction: $Cr_2O_7{}^{2-} + 14\ H^+ + 6\ e^- \rightarrow 2\ Cr^{3+} + 7\ H_2O$

Net reaction: $6\ Fe^{2+} + Cr_2O_7{}^{2-} + 14\ H^+ \rightarrow 6\ Fe^{3+} + 2\ Cr^{3+} + 7\ H_2O$

9) There are 6 Fe, 2 Cr, 7 O and 14 H on each side of the net equation and the total overall charge on each side is +24, so the net equation is balanced.

Exs. 21.2 and 21.3 in the text illustrate other examples of balancing equations for redox reactions occurring in acidic solution. **Pay particular attention to Step 7 in the solutions for these examples, because they illustrate canceling common reactants and products.**

Balancing Equations for Oxidation-Reduction Reactions Occurring in Basic Solutions:

In basic solutions, OH^- and H_2O are able to participate in the reaction in addition to those substances given in the skeleton equation. Thus, we can

> • add 2 OH^- ions for each deficiency of 1 O atom to the side of each half-reaction that is deficient in O atoms and simultaneously add 1 H_2O to the opposite side for every 2 OH^- ions added to maintain an O and H atom balance, as shown for the tin oxidation half-reaction in Step 4 in the solution for Ex. 21.4 in the text

OR

> • follow the entire procedure given for acidic solutions and then neutralize any H^+ ions appearing in the equation for the overall net reaction. This can be done by
>
> > • adding one OH^- for each H^+ present to each side of the equation,
> >
> > • allowing the OH^- and H^+ to combine to form H_2O, and
> >
> > • cancelling some H_2O, if H_2O appears on both sides of the equation following the neutralization combination.

For example, suppose the net equation obtained by using the acidic solution method has the form

$$\ldots\ldots + 3\ H_2O \rightarrow \ldots\ldots + 6\ H^+$$

and the reaction is known to be taking place in basic solution. We would add 6 OH^- ions to both sides of the net equation to provide 6 OH^- to neutralize the 6 H^+ ions and to also maintain the atom and charge balance of the net equation,

$$\ldots\ldots + 3\ H_2O + 6\ OH^- \rightarrow \ldots\ldots + 6\ H^+ + 6\ OH^-$$

allow the 6 H^+ and 6 OH^- to combine to form 6 H_2O (because we know from Expression 18d that $K = 1.0 \times 10^{14}$ for this reaction),

$$\ldots + 3 \ H_2O + 6 \ OH^- \rightarrow \ldots + 6 \ H_2O$$

and then cancel 3 H_2O from each side of the net equation, giving

$$\ldots 6 \ OH^- \rightarrow \ldots + 3 \ H_2O$$

as the final form of the net equation pertaining to basic solution.

Similarly, the acidic solution net equation form

$$\ldots 8 \ H^+ \rightarrow \ldots + 2 \ H_2O$$

yields the basic solution net equation form

$$\ldots 6 \ H_2O \rightarrow \ldots + 8 \ OH^-$$

and you should try using the procedure outlined above to verify this.

✱ Try practicing balancing equations for redox reactions occurring in acidic and basic solutions until you are comfortable doing them and can do them quickly. Refer to the guidelines given on pp. 284, 285, 287 and 289 as needed, but do try to get to know the method so well that you do not need to continue to refer to them. This should include practicing identifying the elements oxidized and reduced and giving their symbols and ionic charges or oxidation numbers and identifying the respective oxidizing and reducing agents.

21.2 Chemical Change Leading to Electrical Current

A **voltaic cell** is an electrochemical cell which uses an electron transfer reaction to produce an electrical current. The electron transfer that occurs in the reaction between Zn and Cu^{2+} in the test tube shown in Fig. 21.2 in the text occurs in a direct manner, but it can also be made to take place in an indirect manner to produce an electrical current using the voltaic cell set-up shown in Fig. 21.3 in the text.

The Zn metal and Cu^{2+} ions are placed into separate containers called **half-cells;** a piece of zinc metal is suspended in a solution of $ZnSO_4$ in one half-cell and a piece of copper metal is suspended in a solution of $CuSO_4$ in the other and the two are connected by an external wire. The zinc and copper metal pieces act as **electrodes**, which conduct electricity out of and into their respective solutions. In other cases, nonreactive conducting substances, such as platinum wire and graphite, can be used as electrodes.

Electrons flow through the external wire from the zinc electrode, which is called the **anode**, to the copper electrode, which is called the **cathode**. The half-cells are connected by a **salt bridge** that is filled with an inert electrolyte. *The function of the salt bridge is to complete the circuit by the movement of ions and to also maintain electrical neutrality in the half-cells.* In this case, SO_4^{2-} ions leave the salt bridge and move into the anode half-cell on the left as Zn^{2+} ions are formed by oxidation of the Zn anode, and SO_4^{2-} ions move into the salt bridge from the cathode half-cell on the right as Cu^{2+} ions are reduced to Cu at the Cu cathode. In addition, there is some movement of Na^+ and perhaps even Zn^{2+} ions into the cathode half-cell to replace the Cu^{2+} ions that are reduced and plated out as Cu metal.

Oxidation always occurs at anode, and reduction always occurs at the cathode in an electrochemical cell. Thus, the appropriate half-reactions and net cell reaction are given by

Anode:	$Zn(s) \rightarrow Zn^{2+}(aq) + 2\ e^-$
Cathode:	$Cu^{2+}(aq) + 2\ e^- \rightarrow Cu(s)$

Cell reaction: $Zn(s) + Cu^{2+}(aq) \rightarrow Zn^{2+}(aq) + Cu(s)$

Notice the similarity between these equations and those on p. 284.

The terminology pertaining to electrochemical cells functioning as voltaic cells or batteries is summarized in Fig. 21.4 in the text. *Notice that anode is labeled "-", because the oxidation produces electrons that make the electrode negative. Conversely, the cathode is labeled "+", because the reduction consumes electrons, leaving the electrode positive. Also notice that it is customary to show the anode (oxidation) half-cell on the left and the cathode (reduction) half-cell on the right in diagrams of voltaic cells.* ✳ Try preparing a "flash card" duplicating Fig. 21.4 in the text and memorizing it. Also study the Cd/Ni^{2+} voltaic cell in Ex. 21.5 in the text.

21.3 Electrochemical Cells and Potential

Electrons generated at the site of oxidation of a cell (the anode) are thought to be "driven" or "pushed" toward the cathode by an **electromotive force**, an **emf**. *This force is due to the difference in electrical potential energy of an electron at the two electrodes. The movement of electrons through this difference in potential energy can be used to do electrical work.*

The quantity of electrical work done is proportional to the number of electrons that go from the higher to the lower potential energy and the size of the potential difference (difference in potential energy per electrical charge). This leads to

$$\text{Electrical work} = \text{charge} \cdot \text{potential difference} \qquad (21a)$$

Charge is measured in coulombs, where one **coulomb (1 C)** is the quantity of charge that passes a point in an electrical circuit when a current of one ampere flows for one second. The charge on a single electron is 1.6022×10^{-19} C (p. 66 in the text), so there is 1 $C/6.24 \times 10^{18}$ electrons and 9.65×10^4 C/ mol of electrons.

Electrical potential difference is measured in volts, where one **volt** is defined so that one joule of work is performed when one coulomb of charge passes through a potential difference of one volt.

$$\text{Volt} = \frac{1 \text{ joule}}{1 \text{ coulomb}} \quad \text{or} \quad \text{Joule} = 1 \text{ volt} \cdot 1 \text{ coulomb} \qquad (21b)$$

The maximum amount of electrical work, measured in J, that can be accomplished by a voltaic cell is therefore equal to the product of the cell potential, measured in volts, times the charge passing through the circuit. The cell potential depends on the substances making up the cell and their pressures or concentrations. The quantity of charge depends on the nature of the reactants and the quantity of reactants consumed.

Because cell potentials depend on the pressures and concentrations of the substances involved in the cell reaction, **standard conditions** have been defined for electrochemical measurements. They are the same as were adopted in Chapter 20 for thermodynamics (p. 274), because Expression 21c tells us thermodynamics and cell potentials are related. That is, *reactants and products in solution must be present at a concentration of 1.0 M, gases must be present at a pressure of 1.0 bar and all must be present at 298 K, unless stated otherwise.* Cell potentials measured under these conditions are called **standard cell potentials** and are designated E°, where the E comes from electromotive force. *Cell potentials for product-favored electrochemical reactions are positive*, by definition. For example, E° for the Zn/Cu^{2+} cell of Fig. 21.3 in the text is +1.10 V.

E° and ΔG°_{rxn}

The standard potential E° is a quantitative measure of the tendency of the reactants to proceed to products under standard conditions and so is the standard free energy change for a reaction, ΔG°_{rxn}. The relation between them is

$$\Delta G^{\circ}_{rxn} = -nFE^{\circ} \qquad (21c, 21.1)$$

where n is the number of moles of electrons transferred between the oxidizing agents and reducing agents in the balanced redox equation and F is the Faraday constant, 9.6485309×10^4 J/V·mol electrons, taken to be 9.65×10^4 J/V·mol electrons. *The minus sign appears in this equation because product-favored electrochemical reactions have negative free energy changes and positive cell potentials.* Ex. 21.6 in the text involves a calculation using Expression 21c.

Calculating the Potential E° of an Electrochemical Cell

An oxidation-reduction reaction is the sum of an oxidation half-reaction and a reduction half-reaction suggesting the potential is also the sum of an oxidation potential and a reduction potential. The problem is that potentials for isolated half-reactions cannot obtained directly because the potential measures the potential energy difference in two different chemical environments. However, it is always possible to measure the standard potential for a half-reaction in combination with another half-reaction, so it is possible to select one half-reaction to be the standard against which all other half-reactions are measured. The half-reaction that was chosen to be this standard is the **standard hydrogen electrode (SHE)**,

$$2 \text{ H}_3\text{O}^+(\text{aq}, 1 \text{ M}) + 2 \text{ e}^- \rightarrow \text{ H}_2(\text{g}, 1 \text{ atm}) + 2 \text{ H}_2\text{O}(\ell) \qquad E^{\circ} \equiv 0.0 \text{ V}$$

$$(21d)$$

and it is assigned a potential of exactly 0.0 V.

The potential for the H_3O^+/H_2 reaction is zero in either direction, so when this half-reaction is used in combination with any other half-reaction in an electrochemical cell, the measured cell potential is assigned to the other half-reaction. For example, the standard cell potential that is measured for the oxidation of Zn by H_3O^+ (Fig. 21.7 in the text) is +0.76 V, so this is the potential that is assigned to the oxidation of Zn to Zn^{2+} under standard conditions. Similarly, the standard cell potential that is measured for the reduction of Cu^{2+} ions by H_2 (Fig. 21.8 in the text) is +0.34 V, so this is the potential that is assigned to the reduction of Cu^{2+} to Cu under standard conditions. Then applying these potentials to the Zn/Cu^{2+}

reaction under standard conditions gives:

Anode, oxidation: $Zn(s) \rightarrow Zn^{2+}(aq,1M) + 2\ e^-$
Cathode, reduction: $Cu^{2+}(aq,1M) + 2\ e^- \rightarrow Cu(s)$

Net reaction: $Zn(s) + Cu^{2+}(aq,1M) \rightarrow Zn^{2+}(aq,1M) + Cu(s)$

with $E^{\circ}_{net} = E^{\circ}\ Zn/Zn^{2+} + E^{\circ}\ Cu^{2+}/Cu = +0.76\ V + +0.34\ V = 1.10\ V$,
which is exactly what is obtained when measured experimentally
(Fig. 21.3 in the text). This is an important result, because *it
shows that two half-cell potentials that are measured against
the same standard H_3O^+/H_2 half-cell can be combined to obtain a
result that is equivalent to an experimentally determined value*.
So, the same approach can be used with the Zn/Zn^{2+}, Cu^{2+}/Cu or
H_3O^+/H_2 half-cells and other half-cells to determine the standard
half-cell potentials for other half-cells, as illustrated in Ex.
21.7 in the text and shown in Table 21.1 in the text.

21.4 Using Standard Potentials

*When half-reactions involving elements and their ions are
written in the form*

$$Oxidized\ form + n\ e^- \rightarrow Reduced\ form \qquad (21e)$$

*and listed in order of decreasing E° values, they are listed in
order of decreasing ability to attract electrons and thus act as
oxidizing agents. Conversely, the reduced forms are listed in
order of increasing ability to lose electrons and thus act as
reducing agents.* The situation is analogous to the relative
strengths of members of conjugate acid-base pairs encountered in
Tables 17.3 and 17.4 in the text, and furthermore, *just as
Bronsted acid-base reactions give predominantly the weaker acid-
base pair, so it is with redox reactions which give predominantly the weaker oxidizing agent/reducing agent pair. Thus,
tables like those shown on pp. 969 and 970 in the text can be
used to predict the outcomes of redox reactions under standard
conditions.*

*There are several important points that should be noted concern-
ing tables of standard reduction potentials that are arranged in
order of decreasing E° values:*

  • The strongest oxidizing agents, having the most positive
  standard reduction potentials, are listed at the top left
  whereas the strongest reducing agents, having the most
  positive standard oxidation potentials obtained by changing
  the sign of the associated reduction potentials, are listed
  at the bottom right.

• The reaction of any oxidizing agent on the left with any reducing agent below it on the right leads to a positive reaction potential and is product-favored under standard conditions.

This means that:

  • $F_2(g)$ is the strongest oxidizing agent listed in the table and $F^-(aq)$ is the weakest reducing agent.

  • $Li^+(aq)$ is the weakest oxidizing agent and $Li(s)$ is the strongest reducing agent.

It also means that the $H_3O^+(aq)$ of acids:

  • will react with metals listed below $H_2(g)$ to produce $H_2(g)$ and the corresponding metal ions.

  • will not react with metals listed above $H_2(g)$ to produce $H_2(g)$ and the corresponding metal ions.

✻ Try preparing a "flash card" outlining the general form of Table 21.4; that is, show the top and bottom entries in the table and their standard reduction potentials, label the positions of the strongest and weakest oxidizing agent in the table and the strongest and weakest reducing agent in the table and memorize the information. However, also be prepared to give a rationale for these positions based on the following principle: the more positive the potential, the more thermodynamically favored the process. Also try preparing a second "flash card" summarizing the meaning of the oxidation potentials of metals in terms of reacting with H_3O^+ from acids.

Ex. 21.8 in the text illustrates the use of potentials from Table 21.4 to determine whether reactions are product-favored in the direction written under standard conditions and to calculate the values of E^O_{net} for the same reactions. Notice that this involves using the positions of the reactants in Table 21.4 to make a qualitative prediction and $E^O_{net} = E^O$ oxidation + E^O reduction to calculate E^O_{net} and verify the prediction. **Also notice that changing the stoichiometric coefficients for a half-reaction does not change the value of the E^O. Electrochemical potentials depend on the nature of the reactants and products and their pressures or concentrations but not on the quantities used.**

> **Note:** The volt is defined as "energy/charge." Multiplying a reaction by some number causes both the energy and the charge to be multiplied by the same number. Thus, the ratio "energy/charge = volt" does not change, and this is why electrochemical potentials do not vary with the quantities used.

Ex. 21.9 in the text illustrates the use of potentials from Table 21.1 to determine the direction of an electrochemical reaction, construct an electrochemical cell based on the reaction and calculate the value of E°.

21.5 Electrochemical Cells at Nonstandard Conditions

Oxidation-reduction reactions seldom occur under standard conditions. Even if they were to start at standard conditions, the concentrations and/or pressures would change causing the potential to change. The goal of this section is to consider the equation that allows us to calculate the potential under nonstandard conditions, the Nernst equation.

The Nernst Equation

The relation between the potential under nonstandard conditions, E, the potential under standard conditions, E°, the number of moles of electrons transferred between the oxidizing and reducing agents in the balanced redox equation, n, and the concentrations and/or pressures of the reactants and products is given by the **Nernst equation,**

$$E = E^\circ - (RT/nF)\ln Q \qquad (21f)$$

where Q is the thermodynamic reaction quotient, F is the Faraday constant (9.65×10^4 J/V·mol) and R is the universal gas constant (8.3145 J/K·mol).

At T = 298 K, Expression 21f becomes

$$E = E^\circ - \frac{0.0257 \text{ V}}{n} \ln Q \quad \text{at 298 K} \qquad (21g, 21.2)$$

This expression allows us to calculate the potential produced by a reaction/cell under nonstandard conditions or to determine the concentration or pressure of a reactant or product by measuring the potential produced by the cell when all other concentrations and pressures are known. The first type of calculation is illustrated in Ex. 21.10 in the text.

E^o and the Equilibrium Constant

The cell potential changes as the reaction proceeds and reaches zero volts when the system reaches equilibrium. Thus, when E reaches 0.0 V, the Q term in the Nernst equation is equal to the thermodynamic equilibrium constant K for the reaction. *So, at equilibrium at 298 K, Expression 21g becomes*

$$E = 0 = E^o - \frac{0.0257 \text{ V}}{n} \ln K \quad \text{at } 298 \text{ K} \quad \quad (21h)$$

which rearranges to give

$$\ln K = \frac{n E^o}{0.0257 \text{ V}} \quad \quad \text{at } 298 \text{ K} \quad \quad (21i, 21.3)$$

This is a very useful equation, because it tells us the equilibrium constant for a reaction can be obtained from a calculation or measurement of E^o, as illustrated in Ex. 21.11 in the text.

21.6 Batteries and Fuel Cells

Batteries are specially designed voltaic cells that can be classified as primary and secondary. **Primary batteries** use oxidation-reduction reactions that cannot be reversed easily, so when the reaction reaches equilibrium, the battery is "dead" and is discarded. **Secondary batteries** are often called **storage batteries**, or **rechargeable batteries**. *The reactions in these batteries can be reversed, so the battery can be "recharged."*

The reactants for a battery are contained in a closed container. On the other hand, the reactants for a **fuel cell** are supplied from an outside source.

Primary Batteries

The common dry cell battery, the **zinc-carbon battery**, shown in Fig. 21.9 in the text, was invented in 1866 by Georges Leclanché. *It consists of a zinc container liner anode, a graphite centerpost cathode, and an electrolyte consisting of a moist paste of $ZnCl_2$, NH_4Cl and MnO_2. The reactions taking place at the electrodes and those taking place to remove the gases formed at the cathode can be represented in simplified form as:*

Anode:	$Zn(s) \rightarrow Zn^{2+}(aq) + 2 \text{ e}^-$
Cathode:	$2 \text{ NH}_4^+(aq) + 2 \text{ e}^- \rightarrow 2 \text{ NH}_3(aq) + H_2(g)$
Additional:	$H_2(g) + 2 \text{ MnO}_2(s) \rightarrow H_2O(\ell) + Mn_2O_3(s)$
Additional:	$Zn^{2+}(aq) + 2 \text{ NH}_3(g) + 2 \text{ Cl}^-(aq) \rightarrow Zn(NH_3)_2Cl_2(aq)$

Net reaction: $Zn(s) + 2NH_4Cl(aq) + 2 \text{ MnO}_2(s) \rightarrow H_2O(\ell) + Mn_2O_3(s) + Zn(NH_3)_2Cl_2(aq)$

---

where the last reaction involves the formation of a complex of Zn^{2+} in a Lewis acid-base reaction with NH_3 and Cl^-.

This type of battery delivers 1.5 V, but *if electrical current is drawn rapidly from the battery, the potential drops because the gaseous products of the reduction cannot be removed from the cathode area rapidly enough.* In addition, *the overall shelf-life and useful lifetime of this battery is shortened by the reaction of H_3O^+ ions, produced by the hydrolysis of the weak acid NH_4^+ (salt hydrolysis, 17.8), with the Zn, but it can be prolonged by storing the battery in a refrigerator to slow this reaction. Alkaline batteries do not have this problem and are also better able to deliver the same potential under high current demands, because no gases are formed, and are thus a wise choice to use.*

Other primary batteries include the **mercury** and **lithium batteries** described in the text.

Secondary Batteries

Secondary, or **storage**, batteries can be recharged meaning the original reactant concentrations can be restored by reversing the net cell reaction using an external source of energy. The best known example of a storage battery is the **lead storage battery** used to start vehicles. *Once the vehicle is running, the battery is recharged by current from the vehicle's alternator.*

There are two types of electrodes in a lead storage battery (Fig. 21.12 in the text): one made of porous lead (the reducing agent) and the other of compressed, insoluble lead(IV) oxide (the oxidizing agent). The electrolyte is aqueous sulfuric acid, and the electrode reactions can be represented as:

Anode: $\quad\quad$ $Pb(s) + SO_4^{2-}(aq) \rightarrow PbSO_4(s) + 2 e^-$
Cathode: $\quad$ $PbO_2(s) + 4 H_3O^+(aq) + SO_4^{2-}(aq) + 2 e^- \rightarrow PbSO_4(s) + $
$\quad\quad\quad\quad\quad\quad\quad\quad\quad\quad\quad\quad\quad\quad\quad\quad\quad\quad$ $6 H_2O(\ell)$

Net rxn: $\quad$ $Pb(s) + PbO_2(s) + 2 H_2SO_4(aq) \rightarrow 2 PbSO_4(s) + 6 H_2O(\ell)$

and *the standard cell potential is given by $E_{net}^o = E^o$ oxidation $+ E^o$ reduction $= +0.356$ V $+ 1.685$ V $= +2.041$ V, so a 12 V battery consists of 6 cells connected in series.*

Lead storage batteries produce a relatively low power for their mass and lead is toxic. However, they do produce a relatively constant 2 V and a large initial current, even when cold, which is especially important to people living in cold weather climates. *The voltage remains relatively constant because most of the reactants and products are solids or pure liquids whose concentrations do not change and thus do not affect the voltage, are not included in the Nernst equation. The other advantage is*

---

that the $PbSO_4$ that is formed as soon as Pb^{2+} is formed by the oxidation of Pb and reduction of PbO_2 is insoluble and clings to the electrode plates making it easy to reverse the reactions during recharging. Eventually, after many uses and rechargings, too much $PbSO_4$ leaves the plates and the battery can longer be recharged, so it goes "dead" and is recycled.

Nickel-cadmium (ni-cad) batteries are rechargeable batteries which are suitable for use in small appliances, as described in the text.

Fuel Cells

Fuel cells convert chemical energy provided by a fuel and an oxidant into electrical energy; the reactants are supplied to the cell as needed from an external source. *The best known fuel converts hydrogen and oxygen into water and electricity aboard space flights (See Fig. 21.14 in the text.). The reactions of interest are:*

Anode: $2 H_2(g) + 4 OH^-(aq) \rightarrow 4 H_2O(\ell) + 4 e^-$
Cathode: $O_2(g) + 2 H_2O(\ell) + 4 e^- \rightarrow 4 OH^-(aq)$

Net rxn: $2 H_2(g) + O_2(g) \rightarrow 4 H_2O(\ell)$

The fuel cells on board the Space Shuttle consume 1500 lb H_2 in a typical 7 day mission, generate 190 gal of drinking water and deliver the same power as batteries would that weigh 10 times more.

21.7 Corrosion: Redox Reactions in the Environment

Corrosion refers to the deterioration of metals by product-favored oxidation-reduction reactions. *In the corrosion of iron, the anodic reaction involves the oxidation of iron. The cathodic reaction involves the reduction of O_2, H_2O or H_3O^+ depending on the environment. The overall equation for the corrosion of iron in the presence of both O_2 and H_2O can be written as*

$$4 Fe(s) + 3 O_2(g) + 2 H_2O(\ell) \rightarrow 2 Fe_2O_3 \cdot H_2O(s)$$
red-brown

Chloride ions from sea air or from salt spread on the roads in the winter time accelerate the rate of corrosion. On other hand, the general approaches to inhibiting corrosion include:

   Anodic inhibition. The surface of the iron metal is 1) painted, 2) chemically treated with a Cr(VI) salt to form a Fe_2O_3 and Cr_2O_3 coating that is impervious to O_2 and H_2O or 3) coated with a thin layer of tin, which is harder to oxidize than iron, to form "tin" cans.

Cathodic protection. The surface of the iron metal is coated with or connected to a more active metal that is easier to oxidize and thus makes the iron the cathode at which O_2 is reduced, as illustrated in Fig. 21.18 in the text. The metals commonly used for this are magnesium and zinc, and these metals are said to act as **"sacrificial anodes."** When the iron is coated with zinc, the process is called **galvanizing**. Galvanizing is used with nails, metal garbage cans and chain link fences, among other things, and we soon see the surfaces of such objects dull as a protective coating of insoluble zinc hydroxide forms.

21.8 Electrolysis: Chemical Change From Electrical Energy

Electrolysis uses electrical energy to cause otherwise reactant-favored oxidation-reduction reactions to occur. *The outside source of electricity causes the cathode to be the negative electrode and the anode to be the positive electrode, and this is just the opposite of a voltaic cell which has an internal source of electricity;* see Problem Solving Tips and Ideas 21.2 in the text.

In the case of molten salts, such as $NaCl(\ell)$, anions migrate to the anode and are oxidized whereas cations migrate to the cathode and are reduced. In the case of aqueous salt solutions, there is the additional possibility of oxidizing solvent species at the anode or reducing solvent species at the cathode. In the case of neutral or reasonably neutral solutions, these possibilities are:

Anode: $\qquad 6\ H_2O(\ell) \rightarrow O_2(g)\ +\ 4\ H_3O^+(aq)\ +\ 4\ e^-$

Cathode: $\qquad 2\ H_2O(\ell)\ +\ 2\ e^- \rightarrow H_2(g)\ +\ 2\ OH^-(aq)$

We should expect the oxidation product of H_2O to be oxygen because hydrogen is already in its highest oxidation state in H_2O, and furthermore, we should expect the reduction product of H_2O to be H_2 because oxygen is already in its lowest oxidation state in H_2O.

In practice, we observe that the order of

- oxidation is: I^-, Br^-, Cl^-, H_2O, SO_4^{2-}

• reduction is: transition metal ions, H_2O, 1A, 2A and Al^{3+}

This isn't necessarily the order that would be predicted using standard potentials because the reactions aren't being conducted under standard conditions and complications are caused by the use of high currents, low concentrations and the kinetic phenomenon of overpotential. However, these general orders based on experimental experience can be used to predict the outcomes of most reactions of interest. For example, these orders enable us to predict that the electrolysis of aqueous

• sodium iodide or chloride will usually result in the oxidation of I^- or Cl^- instead of H_2O and the reduction of H_2O instead of Na^+.

• copper(II) chloride will usually result in the oxidation of Cl^- instead of H_2O and the reduction of Cu^{2+} instead of H_2O.

• sodium sulfate will result in the oxidation of H_2O instead of SO_4^{2-} and the reduction of H_2O instead of Na^+, giving the net cell reaction: $2\ H_2O(\ell) \rightarrow 2\ H_2(g) + O_2(g)$.

They furthermore show why objects, such as car parts, can be chrome-plated in aqueous solutions and why aluminum must be produced by the costly electrolysis of molten Al_2O_3 (21.10).

∗ Try preparing a "flash card" showing the orders of oxidation and reduction for aqueous solution species and the half-reactions for the oxidation and reduction of H_2O and memorizing it.

21.9 Counting Electrons

The number of moles of electrons consumed or produced in an electron transfer reaction can be obtained by measuring the current flowing in the external circuit for the given time. The **current** flowing in an electrical circuit is given by the amount of electrical charge that passes a given point per unit of time, and the unit for current is the **ampere**.

Current, I (amps) = $\dfrac{\text{electrical charge (coulombs,C)}}{\text{time (seconds, s)}}$

$$(21j, 21.4)$$

Thus, *the charge that passes through the cell in the given time can be calculated using*

$$coulombs, C = I \text{ (amps)} \cdot time \text{ (s)} \qquad (21k)$$

$$= \frac{coul}{s} \cdot s$$

Then the number of moles of electrons consumed or produced can be calculated by using the Faraday constant, 96,485.21 C/mol or simply 9.65×10^4 C/mol, as a conversion factor.

$$\text{mol electrons} = \text{coulombs, } C \cdot \frac{1 \text{ mol electrons}}{9.65 \times 10^4 \text{ C}} \qquad (21\ell)$$

Finally, the number of moles of electrons can be converted to the number of moles of substance of interest by using the relation between moles of electrons and moles of substance obtained from the half-reaction and the number of grams of substance can be calculated by using molar mass as a conversion factor. The overall logic pattern is

Time (s) $\cdot$ current (amps) $\rightarrow$ Charge (C) $\rightarrow$ Mol e^- $\rightarrow$ (21m)
    Mol substance $\rightarrow$ Mass substance

as illustrated in Ex. 21.13 in the text. On the other hand, Ex. 21.14 in the text uses the reverse pattern to solve for time.

21.10 The Commercial Production of Chemicals by Electrochemical Means

The production of aluminum and chlorine and sodium hydroxide by electrolytic means are described in this Section. You should review these preparations to the extent that they are covered in your lecture course. However, you should recognize that electricity is only used to prepare elements when there are no convenient chemical methods available, because electricity is expensive. *Indeed, it has been determined that producing an aluminum can by recycling only takes 9% of the energy that is required to produce a new can from bauxite, $Al_2O_3 \cdot 2H_2O$, so you should make every effort to recycle aluminum beverage cans.*

21.11 Key Expressions

Standard potential - standard free energy change expression:

$$\Delta G^o_{rxn} = -nFE^o$$

Standard potential- standard half-reaction potential expression:

$$E^o_{net} = E^o \text{ oxidation} + E^o \text{ reduction}$$

***Nonstandard potential - standard potential (Nernst equation)
expressions:***

$$E = E^O - (RT/nF)\ln Q$$

$$E = E^O - \frac{0.0257 \text{ V}}{n} \ln Q \quad \text{at } 298 \text{ K}$$

***Standard potential - thermodynamic equilibrium constant express-
ion:***

$$\ln K = \frac{nE^O}{0.0257 \text{ V}} \quad \text{at } 298 \text{ K}$$

General electrolytic orders of oxidation and reduction:

Oxidation: I^-, Br^-, Cl^-, H_2O, SO_4^{2-}

Reduction: transition metal ions, H_2O, 1A, 2A and Al^{3+}

Electrical charge expression:

coulombs, $C = I$ (amps) $\cdot$ time (s)

21.12 Hints for Answering Questions and Solving Problems

***Identifying oxidation-reduction reactions and balancing equa-
tions for oxidation-reduction reactions:***

The key to being able to classify a reaction as an oxidation-
reduction reaction is being able to assign oxidation numbers
according to the guidelines given on p. 191 in the text and
being able to note changes in oxidation numbers. The key to
being able to correctly balance equations for oxidation-reduction reactions is systematically following the guidelines given
on pages 284, 285, 287 and 289 and making sure you write the
formulas and charges of all the species correctly in each step
of the balancing process.

> **Note:** The neutralization of H^+ method of balancing
> equations for oxidation-reduction reactions occur-
> ring in basic solution is highly recommended.

***Qualitatively predicting whether an oxidation-reduction will be
product-favored under standard conditions:***

The general principle concerning the predominant direction of an
oxidation-reduction reaction is that: ***the stronger oxidizing
agent + the stronger reducing agent gives predominantly the
weaker oxidizing agent + the weaker reducing agent***. Thus, the

---

reaction of any oxidizing agent on the left of Table 21.1 with any reducing agent below it on the right side of Table 21.1 is predicted to be product-favored under standard conditions.

Calculating the standard potential for an oxidation-reduction reaction and predicting whether it will be product-favored under standard conditions:

The equation $E^{\circ}_{net} = E^{\circ}$ oxidation + E° reduction can be used to calculate the standard potential for an oxidation-reduction reaction, and if it is a positive value, the reaction is predicted to be product-favored in the direction written under standard conditions. If it is a negative value, the reaction is predicted to be reactant-favored in the direction written and thus product-favored in the reverse direction under standard conditions.

The key to correctly calculating E°_{net} is to remember that

- E° oxidation is obtained by reversing the sign of the corresponding reduction potential.

- E° does not change with the number of times the process takes place and is therefore not multiplied when the half-reaction is multiplied.

Using the Nernst equation:

The Q that is used in the Nernst equation is the thermodynamic reaction quotient. This means compounds and ions in aqueous solution are represented by their molar concentrations, which are assumed to be nearly identical to their molal concentrations, and gases are represented by pressures expressed in bars.

Determining the value of n to be used to calculate free energy changes, nonstandard potentials and thermodynamic equilibrium constants:

n represents the number of electrons lost by the reducing agent and the number of electrons gained by the oxidizing agent in a balanced equation. **The value of n** can be determined from the half-reactions combined whenever half-reactions are used to obtain the net equation; it **is the number of electrons that cancels out when the half-reactions are combined. If the half-reactions are not given, the value of n can be determined from changes in oxidation numbers of elements during net equations.**

Conducting calculations for electrolysis reactions:

The operations involved in carrying out the calculations for Ex. 21.13 in the text can be combined to give

$$g = \text{time(s)} \cdot \text{current(C/s)} \cdot \frac{1 \text{ mol } e^-}{9.65 \times 10^4 \text{ C}} \cdot \frac{1 \text{ mol substance}}{n \text{ mol } e^-} \cdot \frac{M \text{ g}}{1 \text{ mol}}$$

which can be solved for g, time, current, n or M when all but one item in this equation is given in the stated problem.

CHAPTER 22

THE CHEMISTRY OF THE
MAIN GROUP ELEMENTS

The elements of the A groups of the periodic table are often called the **main group elements.** According to Fig. 22.1 in the Chapter opening in the text, these elements are more abundant in the earth's crust than are the transition metals. In addition, according the list of the top chemicals produced in the United States that is given on the inside back cover of the text, the top 10 chemicals are main group elements or their compounds. This is why this chapter is devoted to describing the chemistry of selected main group elements.

22.1 The Periodic Table - A Guide to the Elements

We have seen remarkable similarities in the properties of the elements of given groups in the periodic table in Chapters 2, 3, 8, 9 and 10 in the text. Table 22.1 in the text also shows similarities in the common hydrides, common oxides and highest oxidation states of members of the A groups of elements.

We have also noted trends in metallic character in the periodic table. *Metallic character increases as we move to the left in any period below the first period and increases as we move down a group* (Fig. 2.14 in the text). The trend within groups is particularly evident in Group 4A where carbon is a nonmetal, silicon and germanium are metalloids and tin and lead are metals (Fig. 22.2 in the text).

Valence electrons

The main group elements (A group elements) have ns and np valence electrons (8.4). The noble gases of Group 8 all have completed electron subshells; helium has the electron configuration $1s^2$, and the others have ns^2np^6 octets of valence electrons. The first three noble gases, He, Ne and Ar, are chemically unreactive and the last two Kr and Xe are only slightly reactive (8.7). The observation that other elements often react to achieve a noble gas configuration can be used to predict their chemical behavior.

Ionic Compounds of Main Group Elements

The metals of Groups 1A and 2A form M^+ and M^{2+} cations having noble gas configurations, and the nonmetals of Groups 6A and 7A form X^{2-} and X^- anions having noble gas configurations. In addition, aluminum forms Al^{3+} having a noble gas configuration, and nitrogen is known to form a N^{3-} anion having a noble gas configuration (3.3). **These elements react with one another in reactions in which the metals are oxidized and the nonmetals are reduced and the product is a crystalline ionic solid having a high melting point. Ionic compounds are usually nonconductors of electricity in the solid state but are conductors in the molten state or when dissolved in water** (3.4, 8.7).

Ex. 22.1 in the text illustrates the use of the periodic table to predict the formulas of ionic compounds and to write equations for their formation. You can usually refer to Fig. 3.10 to verify your predictions when asked to complete problems of this type.

Covalent Compounds and Electron Configurations

Involving all valence electrons in the formation of compounds is a frequent and reasonable occurrence in main group chemistry. Thus, the halogens, especially fluorine, often cause other elements to exist in their highest possible oxidation states. This observation can be used to predict the formulas of covalent compounds formed between two nonmetals, but you will also want to review the list of covalently bonded polyatomic ions given in Table 3.2 in the text and the naming of covalent compounds described on pp. 128-130 in the text when working in this section.

22.2 Hydrogen

Hydrogen accounts for 0.9% by mass and is ninth in abundance in the earth's crust, where it occurs primarily in water and in fossil fuels. It was used as a fuel and was later used in a fuel mixture with carbon monoxide called **water gas** or **syngas**. However, carbon monoxide is highly toxic, the heat produced is only about one-half that from the combustion of coal gas and the flame is invisible, so it is no longer used as a fuel.

Synthesis of Hydrogen Gas

About 300 billion L (STP) of hydrogen gas is produced annually worldwide. **The largest quantity is produced by the catalytic steam re-formation of hydrocarbons. This process uses methane, chief component of natural gas, as the primary starting material, and the reaction**

$$CH_4(g) + H_2O(g) \rightarrow 3\ H_2(g) + CO(g) \quad \Delta H = +206\ kJ \quad (22a)$$

---

is rapid between 900°C to 1,000°C and goes nearly to completion.
In the second step, known as the **water gas shift reaction**, the
CO formed reacts with more $H_2O(g)$ at 400°C to 500°C giving

$$CO(g) + H_2O(g) \rightarrow H_2(g) + CO_2(g) \qquad \Delta H = -41 \text{ kJ} \qquad (22b)$$

The $CO_2(g)$ is removed by reaction with CaO(s) to form $CaCO_3(s)$.

Common laboratory methods of producing hydrogen are given in
Table 22.3 in the text. * Try preparing separate "flash cards" summarizing
the commercial catalytic steam re-formation of hydrocarbons method of producing
hydrogen gas and the laboratory methods of producing hydrogen gas and memorizing
them.

Properties of Hydrogen

Hydrogen is a colorless gas under standard conditions and is the
least dense gas known. Its very low boiling point of 20.7 K,
reflect its nonpolar character and low molecular mass, low
dispersion forces of attraction.

*Hydrogen combines chemically with virtually every other element
except the noble gases and forms three different types of binary
compounds.*

Ionic Metal Hydrides:

Ionic metal hydrides formed by direct combination with Group 1A
and 2A metals contain the hydride ion, H^-, in which the oxidation
number of hydrogen is -1.

Covalent Hydrides

Covalent hydrides are formed with the nonmetals, and hydrogen is
in the -1 oxidation state when combined with the more electronegative elements carbon, nitrogen, oxygen, sulfur, selenium,
fluorine, chlorine, bromine and iodine (Fig. 9.7 in the text).

Interstitial Hydrides

Hydrogen is absorbed by many metals forming **interstitial
hydrides,** in which hydrogen atoms reside in the spaces (interst-
ices) between metal atoms in the crystal lattice. Most are
nonstoichiometric compounds and release hydrogen gas when
heated.

* Try preparing a "flash card" summarizing the characteristics of these three types
of hydrides and memorizing it.

Some Uses of Hydrogen

The largest use of H_2 is in the production of ammonia, NH_3, by the Haber process (16.3). A large amount is also used to make methanol, CH_3OH (p. 513 in the text). Methanol is used as a fuel additive to cause gasoline to burn more cleanly and also prevents "fuel line freeze" by dissolving traces of water that often contaminate gasoline.

22.3 Sodium and Potassium

Sodium and potassium are the fifth and sixth most abundant elements in the earth's crust at 2.6% and 2.4% by mass, respectively. They are highly reactive with oxygen, water and other oxidizing agents (Fig. 2.17 in the text), and exist as 1+ ions in their compounds. Their compounds are water soluble (Fig. 4.7 in the text), so they exist primarily as dissolved salts in the waters of the earth or in underground deposits that are the residue of ancient seas (p. 169 in the text).

The solubilities of NaCl and KCl are comparable, but ground waters contain more NaCl than KCl. This is because potassium is an important element for plant growth and is preferentially taken up by plants.

Preparation and Properties of Sodium and Potassium

The pure metals were first prepared by the English scientist Sir Humphry Davy in 1807 by electrolyzing the molten carbonates Na_2CO_3 and K_2CO_3. Sodium is now produced by electrolysis of molten NaCl (p. 1012 in the text), and potassium is now produced by the reaction of sodium vapor with molten KCl.

$$Na(g) + KCl(\ell) \rightarrow K(g) + NaCl(\ell) \qquad (22c)$$

The equilibrium constant for the reaction is less than 1, but the reaction is driven to completion by the continuous removal of the potassium vapor.

Both sodium and potassium are silvery metals that are soft and easily cut with a knife (Fig. 2.14 in the text). Their melting points are quite low, 93.5°C for Na and 65.65°C for K. Their densities are also low, just a bit less than the density of water. They react quickly with air and water and must be stored under kerosene or mineral oil.

They are called the **alkali metals** because they react with water to produce an aqueous solution of the metal hydroxide and hydrogen (p. 7 and Fig. 2.17 in the text). They react with the halogens to give the expected MX salts, but the primary product of the reaction of sodium with oxygen is sodium **peroxide**, Na_2O_2, and that of potassium is potassium **superoxide**, KO_2. The latter is used in confined areas, such as aircraft, space craft and submarines, for the removal of CO_2 and the replenishment of O_2, because the reaction generates a greater volume of O_2 than the volume of CO_2 taken in.

$$4\ KO_2(s)\ +\ 2\ CO_2(g)\ \rightarrow\ 2\ K_2CO_3(s)\ +\ 3\ O_2(g) \quad (22d)$$

Sodium compounds of Commercial Interest

Electrolysis of aqueous sodium chloride (Fig 21.23 in the text) produces chlorine and sodium hydroxide.

$$2\ NaCl(aq)\ +\ 2\ H_2O(\ell)\ \rightarrow\ Cl_2(g)\ +\ 2\ NaOH(aq)\ +\ H_2(g) \quad (22e)$$

In 1993, 11.1 billion kg of Cl_2 and 11.7 billion kg of NaOH were produced in the United States by this process, making the so-called chlor-alkali industry one of our largest chemical industries.

Other important compounds of sodium include: Na_2CO_3 (**soda ash**), used in making glass, water treatment, pulp and paper manufacture, and in cleaning materials; $NaHCO_3$ (**baking soda**) used in cooking and added to salt to convert traces of hygroscopic $MgCl_2$ to nonhygroscopic $MgCO_3$ so salt will not clump together.

22.4 Calcium and Magnesium

The Group 2A elements are called the **alkaline earths** because their compounds have high melting points, and the alchemists of medieval times considered any solid substance that did not melt or was not turned into another compound by fire to be an "earth." CaO, for example, melts at 2572°C, which is well beyond the range of an ordinary fire. *The high melting points of the alkaline earth compounds are due to the strong interactions between the M^{2+} cations and the anions present* (p. 119 in the text).

Similar to 1A elements, 2A elements are quite reactive and are only found in chemical combination with other elements. However, *the 1A compounds are water soluble whereas many of the 2A compounds are not owing to stronger forces of attraction between anions and M^{2+} cations than M^+ cations*. This explains the common occurrence of **limestone** ($CaCO_3$), **gypsum** ($CaSO_4 \cdot H_2O$), fluorspar (CaF_2), magnesite ($MgCO_3$) and dolomite ($MgCO_3 \cdot CaCO_3$). Marble is fairly pure calcite formed by the crystallization of $CaCO_3$ under high pressure. Another form of calcite is Icelanidic spar, which forms large, clear crystals (Fig. 2.18 in the text).

Properties of Calcium and Magnesium

Calcium and magnesium are fairly high-melting, silvery metals. They form MX_2 compounds with the halogens, MO and MS compounds with oxygen and sulfur, react with water to form the $M(OH)_2$ hydroxide and H_2 and lastly react with acids to form H_2 and a salt of the metal and the anion of the acid.

Metallurgy of Magnesium

The metallurgy of magnesium is of interest, because **magnesium is obtained from seawater and a key step in recovering it takes advantage of the stronger lattice and consequently lower solubility of $Mg(OH)_2$ compared to $Ca(OH)_2$.** The metallurgy and uses of magnesium are described in detail in the text, but it is worth noting that the magnesium metal is produced by the electrolysis of molten magnesium chloride. **Electrolysis is commonly used to prepare the metals of 1A, 2A and Al, because they are active metals and their ions are difficult to reduce.** ∗ Try preparing a "flash card" summarizing the chemical equations for the production of magnesium from seawater and memorizing it.

Calcium Compounds of Commercial Interest

Fluorspar, CaF_2, is used in the steel industry where it is added to the mixture of materials that is melted to make crude iron. Its function is to remove some impurities and improve the separation of the molten metal from the layer of silicate impurities and byproducts that comes from reducing iron ore to the metal, the slag (Chapter 23 in the text). **It is also used to manufacture hydrofluoric acid by reaction with concentrated sulfuric acid**

$$CaF_2(s) + H_2SO_4(\ell) \rightarrow 2\ HF(g) + CaSO_4(s) \qquad (22f)$$

where the driving force is the formation of the weaker and more volatile acid, HF. Hydrofluoric acid reacts with the SiO_2 of glass and therefore cannot be kept in glass containers. It is used to make cryolite, Na_3AlF_6, which is used in making aluminum and fluorocarbons, such as polytetrafluoroethylene, which is sold as Teflon. It is also used in the conversion of crude oil to gasoline.

Calcium fluoride also occurs in combination with calcium phosphate in fluorapatite, $CaF_2 \cdot 3Ca_3(PO_4)_2$. Apatites are collectively called **phosphate rock** and most of the more than 100 million tons of phosphate rock mined annually is converted to phosphoric acid by reaction with sulfuric acid.

$$CaF_2 \cdot 3Ca_3(PO_4)_2(s) + 10\ H_2SO_4(aq) \rightarrow 10\ CaSO_4(s) + 6\ H_3PO_4(aq) + 2\ HF(g)$$

(22g)

Phosphoric acid is used in cola drinks and the manufacture of a multitude of products, including fertilizers and detergents.

Limestone, $CaCO_3$, has been used to make statutes and buildings. It often contains traces of $MgCO_3$ and has been used in agriculture to neutralize acidic compounds in the soil and supply the essential nutrients Ca^{2+} and Mg^{2+}. More recently, it has been used for "liming" lakes which have become too acidified by the effects of acid rain.

The thermal decomposition of limestone to **lime**, CaO, is important to the steel industry (Chapter 23), and lime has been used as a paste of lime, sand and water as mortar for several thousand years. The lime and water combine to give $Ca(OH)_2$ in a process that is called "slaking," so the $Ca(OH)_2$ product is called "**slaked lime**." The mortar slowly absorbs CO_2 from the air converting the slaked lime to calcium carbonate and the grains of sand are bound together by the particles of calcium carbonate resulting in a hard material.

$$Ca(OH)_2(s) + CO_2(g) \rightarrow CaCO_3(s) + H_2O(\ell)$$

(22h)

"**Hard water**" contains primarily Ca^{2+} and Mg^{2+} ions. Water containing dissolved CO_2 reacts with limestone in the reversible reaction

$$CaCO_3(s) + CO_2(g) \rightleftharpoons Ca^{2+}(aq) + 2\ HCO_3^-(aq)$$

(22i)

that is responsible for the formation of limestone caves with stalactites and stalagmites, as well as hard water. **When hard water is heated, the solubility of CO_2 decreases and the equilibrium shifts to the left forming limestone. This is the cause of coatings of $CaCO_3$ on cookware and of the clogging of pipes used with hot water heating.**

When dry ice, $CO_2(s)$, is added to a solution of $Ca(OH)_2$, the reaction first turns cloudy with the formation of $CaCO_3$ and then turns clear as the $CaCO_3$ dissolves because of the formation of Ca^{2+} and HCO_3^-. Thus, Expressions 22g and 22h account for the observations presented in the Chapter-opening Chemical Puzzler.

22.5 Aluminum

Aluminum is the third most abundant element in the earth's crust at 7.4% by mass. **Pure aluminum is soft and weak and loses strength rapidly above 300°C. It is often alloyed with small amounts of other metals to improve its properties.** A typical alloy contains 4% Cu with smaller amounts of Si, Mg, and Mn. However, the softer, more corrosion resistant alloy that is used to make window frames, overhead highway sign holders and cooking utensils contains just Mn.

Aluminum is easily oxidized but forms a thin, tough and transparent protective coating of Al_2O_3. If this coating is scratched, it rapidly self-repairs by forming a new coating.

Metallurgy of Aluminum

Aluminum is found in clays called aluminosilicates, which weather and break down to give bauxite, $Al_2O_3 \cdot n$ H_2O. **Bauxite is separated from Fe_2O_3 and SiO_2 impurities by using the acid-base chemistry of the Bayer process. SiO_2 is an acidic oxide, Al_2O_3 is an amphoteric oxide and Fe_2O_3 is a basic oxide, so the first two dissolve in hot concentrated caustic soda (NaOH), leaving Fe_2O_3 to be filtered off.**

$$Al_2O_3(s) + 2 \text{ NaOH(aq)} + 3 \text{ H}_2O(\ell) \rightarrow 2 \text{ Na}[Al(OH)_4](aq) \quad (22j)$$

$$SiO_2(s) + 2 \text{ NaOH(aq)} + 2 \text{ H}_2O(\ell) \rightarrow Na_2[Si(OH)_6](aq) \quad (22k)$$

By treating the solution containing the aluminum and silicon anions with CO_2, Al_2O_3 is reprecipitated. The reaction of CO_2 with water gives H_2CO_3, which is a weak acid and reacts with the $Al(OH)_4^-$ to give Al_2O_3.

$$H_2CO_3(aq) + 2 \text{ Na}[Al(OH)_4](aq) \rightarrow Na_2CO_3(aq) + Al_2O_3(s) + 5 \text{ H}_2O(\ell)$$

$$(22\ell)$$

The aluminum is then obtained by the electrolysis of molten Al_2O_3 using cryolite, $NaAlF_6$, to lower the melting point (Fig. 21.22 in the text).

✳ Try preparing a "flash card" summarizing the chemistry and chemical equations for the preparation of aluminum, including the electrolysis reaction, and memorizing it.

Properties of Aluminum and Its Compounds

Aluminum dissolves in hydrochloric acid, HCl(aq), as it would be expected to do based on its position in Table 21.1. However, it fails to dissolve in nitric acid, HNO_3(aq), which provides the oxygen atoms needed for forming a protective coating of Al_2O_3.

Soluble aluminum salts dissolve in water to give acidic solutions due to hydrolysis of $Al(H_2O)_6^{3+}$ (17y). On the other hand, the lattice interactions of Al_2O_3 are so strong that it is insoluble in water and is extremely hard. In the crystalline form, it is known as *corundum* and is used as the abrasive in grinding wheels, "sandpaper" and toothpaste.

Some gems are impure Al_2O_3. Rubies, for example, contain some Cr^{3+} in place of Al^{3+}, and blue sapphires contain some Fe^{2+} and Ti^{4+} in place of Al^{3+}.

22.6 Silicon

Silicon is the second most abundant element in the earth's crust. Silicon-based natural materials were used in the making of pottery at least 6,000 years ago in the Middle East, and silicon-based semiconductors have enabled the computer revolution of the past two decades to take place.

Reasonably pure silicon is made in large quantities by heating pure silica sand with purified carbon, as coke, to approximately 3,000°C in an electric furnace.

$$SiO_2(s) + 2\ C(s) \rightarrow Si(\ell) + 2\ CO(g) \qquad (22m)$$

The molten Si is drawn off the bottom of the furnace and allowed to cool to a shiny blue-gray solid (Fig. 2.14 and p. 486 in the text). The crude Si is converted to high-purity Si for use in the electronics industry by converting it to volatile $SiCl_4$, purifying the $SiCl_4$ by distillation, reducing the purified $SiCl_4$ to Si using Mg and purifying the Si by **zone refining** (Fig. 22.6 in the text). ✳ Try preparing a "flash card" summarizing the procedure for preparing high-purity Si, including the chemical equations for the reactions, and memorizing it.

Silicon Dioxide

The simplest oxide of silicon is SiO_2, commonly called **silica**. Silica is a major constituent of many rocks, such as granite and sandstone. **Quartz** is a pure crystalline form of silica. The introduction of impurities into quartz produces gemstones, such as amethyst (Fig. 13.36 in the text).

SiO₂ is a network solid, in which each Si is covalently bonded to four oxygen atoms (p. 1020 in the text), and has a melting point of 1610°C as quartz. On the other hand CO₂ exists as individual nonpolar molecules in which each O atom is bonded to the central C atom by the use of double bonds involving p-p π overlaps (pp. 470 and 471 in the text), and exists as a gas at temperatures as low as -78.5°C at 1 atm (p. 644 in the text). Why the difference? Atoms of the third period and below are presumably too large to form effective p-p π overlaps, because their sigma bonds are too long, and this favors the formation of additional single bonds to achieve a noble gas configuration; the exception being the smaller S atom that is known to form multiple bonds with C, N and O but not itself.

Crystalline quartz is used to control the frequency of almost all radio and television transmissions. In addition, amorphous SiO_2, called silica gel, is used as drying agent because it can absorb up to 40% of its own weight of water. Small packets of silica gel are often placed in boxes of merchandise for storing and shipping.

The Silicate Minerals

The **silica minerals** are built from tetrahedral SiO_4 units and include asbestos, mica, clays, feldspars and zeolites. The structures of these minerals are difficult to describe in the absence of illustration, so you are referred to pp. 1021-1023 in the text for your study of this material.

Silicone Polymers

Si can react with methyl chloride, CH_3Cl, to give $(CH_3)_2SiCl_2$ which reacts with water to produce $(CH_3)_2Si(OH)_2$ plus HCl. The reaction of n $(CH_3)_2Si(OH)_2$ monomer units with themselves then gives a polymer chain n units long.

$$n \ (CH_3)_2Si(OH)_2 \rightarrow \ -[(CH_3)_2SiO]_n - \ + \ n \ H_2O \qquad (22n)$$

The polydimethylsiloxane product formed by this reaction involves a chain of alternating silicon and oxygen atoms, as shown on p. 1024 in the text. Silicones are nontoxic, have good stability to heat, light and oxygen, are chemically inert and have valuable antistick and antifoam properties. They are used in a wide variety of products: as lubricants, as the antistick material for peel-off labels, in lipstick, suntan lotion and car polish, and as the antifoam substance in stomach remedies. ✱ Try preparing a "flash card" showing the chemical equations for the conversion of Si to polydimethylsiloxane and memorizing it.

22.7 Nitrogen and Phosphorus

Nitrogen and phosphorus are not among the 10 most common elements in the earth's crust, but they are certainly essential to life on planer earth. Nitrogen and its compounds, such as ammonia, nitric acid, ammonium nitrate, and urea, play a key role in our economy, and we have noted that phosphoric acid is used in the production of a multitude of compounds. The major use of all these compounds is in fertilizers.

Nitrogen is found primarily as N_2 in the atmosphere (78.1% by volume and 75.5% by weight). Phosphorus occurs primarily in phosphate minerals that contain the tetrahedral PO_4^{3-} ion or a derivative of it and are therefore called **orthophosphates.** The largest source of phosphorus is the **apatite** mineral family, which have the general formula 3 $Ca_3(PO_4)_2 \cdot CaX_2$ (X= F, Cl or OH). Nitrogen and phosphorus are part of every living organism and constitute about 3% and 1.2% by weight, respectively, of the human body. Nitrogen occurs in proteins and nucleic acids, and phosphorus occurs in nucleic acids, phospholipids and **hydroxyapatite**, which is an important constituent of our bones and tooth enamel. *Fluoridation involves substituting the less basic F^- ion for OH^-, making tooth enamel less susceptible to attack by acids formed by the action of bacteria on sugars and thus prevents tooth decay.*

The Elements: Nitrogen and Phosphorus

Elemental nitrogen, N_2, is a colorless gas that liquifies at 77K (-196°C). *Its strong triple bond (945.4 kJ/mol) causes its reactions to have high activation energies, so it is unreactive (15.5). However, it does react with Li and Mg to form nitrides containing the N^{3-} ion.*

Elemental phosphorus exists as tetrameric P_4 molecules (Fig. 3.3 in the text), and this is still another example of the failure of elements of the third period and below to form effective p-p π overlaps. It is produced in large quantities by the reduction of phosphate minerals.

$$2\ Ca_3(PO_4)_2(s)\ +\ 10\ C(s)\ +\ 6\ SiO_2(s) \rightarrow P_4(s)\ +\ 6\ CaSiO_3(s)\ +\ 10\ CO_2(g) \qquad (22o)$$

Nitrogen Compounds

One of the most interesting features of nitrogen chemistry is that it is known to exist every oxidation state between -3 and +5. This is the maximum range for a second-period element.

Ammonia and Nitrogen Fixation:

Nitrogen gas, N_2, cannot be used by plants until it is "fixed," that is converted into a form that can be used by living systems. Nitrogen fixation occurs naturally with organisms such as blue algae and with a few field crops such as alfalfa and soybeans, which contain nitrogen fixing bacteria. However, most new varieties of wheat, corn and rice that grow fast or have been bred to provide high levels of protein require an outside source of nitrogen. The Haber process for producing ammonia, NH_3, which is used as a fertilizer, was described on p. 778 in the text.

Hydrazine:

Hydrazine, N_2H_4, is a colorless fuming liquid that *acts as weak base, because of the lone pairs of electrons on the nitrogen atoms*, and a strong reducing agent. It is used to remove dissolved oxygen from water used in boilers in large power plants, because the dissolved oxygen could oxidize the metal of the boiler and its pipes causing corrosion.

$$N_2H_4(aq) + O_2(g) \rightarrow N_2(g) + 2\ H_2O(\ell) \tag{22p}$$

Nitrogen Oxides:

The large number of binary nitrogen oxides are shown in Table 22.4 in the text. They are of interest, because all of the positive free energies of formation and are therefore thermodynamically unstable with respect to $N_2(g)$ and $O_2(g)$. However, they are slow to decompose and are said to be kinetically stable.

- **Dinitrogen oxide**: N_2O is a nontoxic, odorless, tasteless gas that is soluble in vegetable fats and is used as a propellant and aerating agent in cans of whipped cream. It is also used as an anesthetic in minor surgery and has come to be known as "laughing gas" because of its effects.

- **Nitrogen monoxide:** NO is a simple odd-electron molecule that has recently been the subject of intense research because it has been found to be important in a number of biochemical processes, and it may be beneficial to newborns with breathing problems. See "NO is no Dud" on p. 414 in the text.

- **Nitrogen dioxide:** NO_2 is the brown gas that is seen when bottles of nitric acid are allowed to stand in sunlight.

$$2\ HNO_3(aq) \rightarrow 2\ NO_2(g) + H_2O(\ell) + \tfrac{1}{2}\ O_2(g) \tag{22q}$$

It is an odd electron molecule that is a culprit in air pollution and is also known to dimerize to give **dinitrogen tetraoxide.**

$$2\ NO_2(g)\ \rightleftharpoons\ N_2O_4(g) \tag{22r}$$

Nitric Acid:

Nitric acid can be formed by the reaction of Chilean saltpeter, $NaNO_3$, with sulfuric acid. However, *most nitric acid is produced from ammonia in the multistep Ostwald process that was described in Section 15.7 as an example industrial catalysis.* Approximately 20% of the NH_3 produced is now used to produce nitric acid by this process. The majority of nitric acid is reacted with ammonia to produce ammonium nitrate for use as a fertilizer.

Nitric acid is a powerful oxidizing agent as indicated by the large, positive E^O values for (22s and 22t)

$$NO_3^-(aq)\ +\ 4\ H_3O^+(aq)\ +\ 3\ e^-\ \rightarrow NO(g)\ +\ 6\ H_2O(\ell) \qquad E^O\ =\ +0.96\ V$$

$$NO_3^-(aq)\ +\ 2\ H_3O^+(aq)\ +\ e^-\ \ \rightarrow NO_2(g)\ +\ 3\ H_2O(\ell) \qquad E^O\ =\ +0.80\ V$$

Thus, nitric acid attacks and oxidizes most metals. Which oxide forms depends on the metal and on the reaction conditions. With copper, NO is formed with dilute acid, and NO_2 is formed with concentrated acid (Fig. 4.18 in the text).

In dilute acid:

$$3\ Cu(s)\ +\ 8\ H_3O^+(aq)\ +\ 2\ NO_3^-(aq)\ \rightarrow\ 3\ Cu^{2+}(aq)\ +\ 12\ H_2O(\ell)\ + \\ 2NO(g)$$
$$\tag{22u}$$

In concentrated acid:

$$Cu(s)\ +\ 4\ H_3O^+(aq)\ +\ 2\ NO_3^-(aq)\ \rightarrow\ Cu^{2+}(aq)\ +\ 6\ H_2O(\ell)\ +\ 2\ NO_2(g)$$
$$\tag{22v}$$

However, a mixture of 1 part by volume concentrated nitric acid to 3 parts by volume concentrated hydrochloric acid, called **aqua regia**, or "*kingly water,*" is needed to dissolve the "noble metals," Au, Pt, Rh and Ir.

22.8 Oxygen and Sulfur

Oxygen is the most abundant element in the earth's crust,

representing just under 50% by weight. It also appears in the atmosphere and the waters of the lakes and seas.

Sulfur is 15th in abundance in the earth's crust. It can also be found in its elemental form in nature, though only in certain concentrated deposits. For example, approximately 10 million tons of sulfur are taken annually from deposits along the Gulf of Mexico that are believed to have been formed by anaerobic ("without air") bacteria feeding on sedimentary sulfate deposits, such as gypsum ($CaSO_4 \cdot 2\ H_2O$). Sulfur also exists in sulfur-containing compounds in oil and natural gas and in metal sulfide ores.

Preparation and Properties of the Elements

Pure oxygen is obtained by the fractional distillation of liquid air, and is third in industrial production in the United States. It can be prepared in the laboratory by the electrolysis of water or the thermal decomposition of $KClO_3$.

$$2\ KClO_3(s) \rightarrow 2\ KCl(s) + 3\ O_2(g) \qquad (22w)$$

At room temperature and pressure oxygen is a colorless gas, but is pale blue when condensed to the liquid at $-183°C$ (Fig. 10.15 in the text). It has two unpaired electrons and is therefore paramagnetic, as discussed in Section 10.2 in the text.

Oxygen also exists as a second, less stable, allotrope called ozone, O_3. It is a blue diamagnetic gas with an odor so strong that it can be detected in concentrations as low as 0.05 ppm. It can be formed by irradiation of O_2 with ultraviolet light, as in sterilizers in beauty shops, or by passing an electric arc through O_2, as during thunderstorms. *We depend on a thin layer of ozone to absorb and thus protect us from ultraviolet radiation from the sun.*

The most important allotrope of sulfur is the yellow, orthorhombic form, which consists of S_8 molecules having the sulfur atoms arranged in a crown-shaped ring (p. 1030 in the text). Sulfur is obtained from underground deposits using the Frasch process (Fig. 22.14 in the text).

The largest use of sulfur is to produce sulfuric acid, the chemical that is formed in larger quantity than any other chemical in the United States. About 70% of the sulfuric acid is used to make "super-phosphate" fertilizer. Sulfuric acid is also used to make iron and steel, petroleum products, synthetic polymers and paper.

The Chemistry of Sulfur

Hydrogen sulfide, H_2S, exists a gas at room temperature and pressure because its intermolecular forces are weaker than those for H_2O. It is toxic but has an odor that can be detected. It has the odor of rotten eggs.

Metal sulfides are heated in air, are roasted, to give either the metal or metal oxide and SO_2. SO_2 is a colorless, toxic gas with a choking odor that dissolves in water, but can be converted to SO_3 to prepare sulfuric acid.

22.9 Chlorine

Chlorine is a yellow-green gas that was first made in 1774 by the Swedish chemist Karl Wilhelm Scheele using a reaction that is still used today for a laboratory source of Cl_2.

$$2\ NaCl(aq) + 2\ H_2SO_4(aq) + MnO_2(s) \rightarrow Na_2SO_4(aq) + MnSO_4(aq) + 2\ H_2O(\ell) + Cl_2(g) \qquad (22x)$$

Chlorine exists as the chloride ion in natural waters and in brine wells. It is 11th in abundance and is the most abundant halogen. Today, chlorine is produced by the electrolysis of brine solutions (Fig. 21.23 in the text) and ranks 8th among industrial chemicals. About 70% of the chlorine produced winds up in organic chemicals like the polyvinylchloride (PVC) plastic products, and another 20% is used to bleach textiles and paper and disinfect water.

Chlorine Compounds

Hydrogen Chloride:

Hydrogen chloride, in the form of hydrochloric acid, ranks 26th among industrial chemicals. It usually prepared by the reaction of NaCl and H_2SO_4.

$$2\ NaCl(s) + H_2SO_4(aq) \rightarrow Na_2SO_4(s) + 2\ HCl(g) \qquad (22y)$$

Gaseous hydrogen chloride has a sharp irritating odor and dissolves in water to give hydrochloric acid, a strong acid. Gaseous hydrogen chloride reacts with metals and metal hydrides to give metal chlorides and hydrogen and with metal oxides to give metal chlorides and water.

Oxoacids of Chlorine:

Oxoacids of chlorine include HOCl, HOClO, $HOClO_2$ and $HOClO_3$ (HClO, $HClO_2$ $HClO_3$ and $HClO_4$), exhibiting the oxidation states of

+1, +3, +5 and +7. ***The first two of these are weak acids, and the last two, having two or more O atoms (non OH O atoms) attached to the central atom are strong acids.*** All are strong oxidizing agents. ✶ Try preparing a "flash card" giving the names of the acids and conjugate base anions listed on p. 1032 in the text and memorizing it.

Cl_2 reacts with water in an oxidation-reduction reaction in which part of the chlorine is oxidized to give hypochlorous acid and the other part is reduced to give the chloride ion.

$$Cl_2(g) + 2\ H_2O(\ell) \rightleftharpoons H_3O^+(aq) + Cl^-(aq) + HOCl(aq) \qquad (22z)$$

When Cl_2 is added to cold aqueous NaOH, the reaction is similar but proceeds much farther to the right giving

$$Cl_2(g) + 2\ OH^-(aq) \rightleftharpoons OCl^-(aq) + Cl^-(aq) + H_2O(\ell) \qquad (22aa)$$

Indeed, the resulting aqueous solution is the "liquid bleach" that is used in home laundries. When calcium hydroxide is used in place of sodium hydroxide, the product is solid $Ca(ClO)_2$ that is sold to sterilize swimming pools.

Perchlorates:

Perchlorate salts, containing ClO_4^-, are the most stable oxochlorine compounds, but they are still powerful oxidizing agents. Pure perchloric acid, $HClO_4$, is a colorless liquid that explodes when shocked and should not be allowed to come into contact with organic compounds because it reacts too violently with them. It even oxidizes silver and gold. However, dilute solutions of the acid are safer to handle because they have less oxidizing power.

Ammonium perchlorate, NH_4ClO_4, bursts into flame when heated above 200°C and produces all gaseous products.

$$2\ NH_4ClO_4(s) \rightarrow N_2(g) + Cl_2(g) + 2\ O_2(g) + 4\ H_2O(g) \qquad (22ab)$$

This is why NH_4ClO_4 is used as the oxidizer in the solid booster rockets for the Space Shuttle along with powdered aluminum as the reducing agent. Each Shuttle launch requires about 750 tons of ammonium perchlorate. It is made by taking advantage of the lower solubility of NH_4ClO_4 in water compared to $NaClO_4$

$$NaClO_4(aq) + NH_4Cl(aq) \rightarrow NaCl(aq) + NH_4ClO_4(s) \qquad (22ac)$$

and more than one-half of the sodium perchlorate currently being manufactured is converted to the ammonium salt.

CHAPTER 23

THE TRANSITION ELEMENTS

The transition metals have an interesting and rich chemistry. Iron, for example, is essential for the transportation of oxygen within us and is also one of the most important substances in our economy because it is used to make various types of steel. Cobalt, on the other hand, is the metal center of Vitamin B_{12}, a compound that acts as a catalyst in the metabolism of carbohydrates, fats and proteins. In addition, the chemistry of the transition metals is of interest because of the variety of colors of their compounds and the variations in their magnetic properties. We will begin with an overview consideration of the properties of the transition elements with emphasis on the first transition series, consider the production of iron and copper and conclude with a consideration of the structures, colors and magnetic properties of coordination compounds.

23.1 Properties of the Transition Elements

Electron Configurations

The chemical behavior of an element is related to its electron configuration, so it is important to know the electron configurations of the d-block elements and their common ions (Fig. 3.10 in the text and Section 8.4). Recall that *the configurations of these metals have the general form [noble gas core]$ns^a(n-1)d^b$. This means valence electrons for the transition elements reside in the ns and (n-1)d subshells* (See Tables 8.2 and 8.4 in the text.).

Oxidation of these elements results in positive ions. In this process, all the s electrons are lost, and, in some cases, one or more of the d electrons are lost as well. The resulting ions have the electron configuration [noble gas core]$(n-1)d^x$. In contrast to ions formed by main group elements, transition metal ions do not have noble gas configurations, and their compounds often possess unpaired electrons, leading to the paramagnetic behavior discussed later in this section.

Oxidation Numbers

Oxidation numbers of +2 and +3 are commonly observed in compounds of the first transition series (Fig. 3.10 in the text). Examples of oxidation reactions of transition metals include those with

- oxygen to form metal oxides,

- halogens to form metal halides, and

- aqueous acids to form hydrated metal cations

(Table 23.1 and Fig. 23.4 in the text).

Recall from Chapter 21 that tables of electrochemical potentials are a source of useful information on oxidations and reductions. *The standard reduction potentials for the first transition series elements are shown in Table 23.2 in the text. All of these metals except copper have positive oxidation potentials and can therefore be oxidized by $H_3O^+(aq)$, because E^o H_3O^+/H_2 = 0.0 V combined with E^o M/M^{2+} or M/M^{3+} = + V gives a positive E^o_{net}.*

The +2 and +3 oxidation states are the most common for the first transition series elements, but they can also be found in a wide range of oxidation states, with the highest oxidation states being observed with the middle elements (Fig. 23.5 in the text). Higher oxidation states are also common in compounds of elements from the second and third transition series, whereas the +2 and +3 ions are encountered less often.

Metal Atom Radii

The radii of transition elements change less rapidly than those of the first three members of a period. They also drop to a minimum around the middle of the series and then rise slowly (Fig. 8.11 in the text). *This variation from the normal trend of decreasing radii across a period can be understood in terms of the $(n-1)d^b$ electron configurations of these elements, though it is actually the location of the outer ns electrons that determines the radii. In the first half of the transition series, the slight increase in effective nuclear charge caused by the simultaneous addition of protons to the nuclei and electrons to the inner $(n-1)d$ orbitals causes the radii to decrease slightly. However, the small increase in radii in the second half of this series of elements is caused by the increase in electron-electron repulsions as the inner d subshell is completed.*

Interestingly, the overall increase in effective nuclear charge that occurs across the fourth-period transition metals causes the radius of gallium to be smaller than that of aluminum instead of larger (Table 23.3 in the text).

Another interesting observation can be made from the graph of Fig. 8.11 in the text and that is that *the radii of the transi-*

tion elements in the fifth and sixth periods are almost identical. The reason for this is that the lanthanide elements are inserted into the periodic table just before the sixth period transition series. The filling of the poor shielding 4f orbitals through the lanthanide elements is accompanied by a steady decrease in size causing the elements of the third transition series to be comparable in size to those of the second transition series. This effect is so significant that it is called the **lanthanide contraction.** The effect of the lanthanide contraction even carries over to the start of the 6p block where the increase in size from In to Tl is less than expected.

Density

A consequence of the variation in metal radii is that the densities of the transition metals first increase and then decrease across a given period (Fig. 23.7 in the text). The change in radii is small but the effect is magnified because density is mass/volume, and volume changes with the cube of the radius ($V = 4/3 \, \pi r^3$ for spheres.).

The lanthanide contraction is the reason that the transition elements in the sixth period have the highest density. The relatively small radii of these elements combined with their larger atomic masses compared to their counterparts of the fifth period causes the sixth-period metal densities to be very large. Indeed, osmium (d = 22.49 g/cm^3) and iridium (d = 22.41 g/cm^3) have the highest densities of all the elements.

Melting Point

The transition metals with the highest melting points occur in the middle of each series (Fig. 23.8 in the text). The bonding between metal atoms was described in terms of the band theory of metals in Section 10.3. However, here it is interesting to note that *the melting point data indicate that the strongest metallic bonds occur when the d subshell is about half-filled, when there is the largest number of unpaired electrons in the isolated atoms. Thus, the d electrons must play an important role in the bonding in these metals*.

Magnetism

The transition elements can have 0 to 5 unpaired electrons. Recall (Section 8.1 in the text) that atoms having unpaired electrons are attracted to magnets and are said to be **paramagnetic.** However, in the cases of iron, cobalt and nickel, the atomic magnets of a group of atoms combine to form a **magnetic domain** in which all of the magnets are oriented in the same direction. The magnitude of the effect is much larger than the effect of paramagnetism and is called **ferromagnetism.** *Ferromag-*

netic materials are unique in that once the atomic magnets are aligned by an external magnetic field, the metal is permanently magnetized. The magnetism can only be eliminated by heating the metal to cause the atoms to vibrate and rearrange the electron spin domains.

23.2 The Commercial Production of Transition Metals

The relatively few minerals from which metals can be extracted profitably are called **ores**, and some of these are listed on p. 169 in the text. *Ores usually consist of the desired mineral and large quantities of impurities,* such as sand and clay, called **gangue** (pronounced "gang"). *A major step in obtaining the metal of interest is separating the mineral from the gangue. The second major step involves converting the ore to the metal.* Pyrometallurgy and hydrometallurgy are the two most common methods used to recover metals from their ores. **Pyrometallurgy** uses high temperatures like those used in the production of iron. **Hydrometallurgy** uses aqueous solutions at relatively low temperatures for the extraction of metals, such as copper, zinc, tungsten and gold.

Iron Production

The production of iron from its ores involves oxidation-reduction reactions carried out in a blast-furnace (Fig. 23.9 in the text). The furnace is charged at the top with a mixture of iron ore (usually hematite, Fe_2O_3), coke (which is primarily carbon), and limestone ($CaCO_3$), and a blast of hot air is forced in at the bottom. The iron(III) is reduced by C and CO, collected as molten iron at the bottom of the furnace and cooled to give impure iron called "**cast iron**" or "**pig iron**." Lime (CaO) formed by the thermal decomposition of the limestone ($CaCO_3$) reacts with the silica (SiO_2) impurity in an acid-base reaction to give a calcium silicate ($CaSiO_3$) "**slag**" that is less dense than iron and forms a floating layer that can be removed.

Pig iron can contain up to 4.5%C, 1.7% Mn, 0.3% P, 0.04% S and 15% Si and is too brittle to be useful, so it must be purified. The most important purification technique is currently the **basic oxygen furnace** technique in which oxygen is blown into molten pig iron to convert the impurities to oxides that escape as gases or react with the basic oxides that are used to line the container. The resulting steel contains about 1.3% C and is known as basic carbon steel. It can be converted to other steels by the addition of other transition elements to produce alloys having specific physical, chemical and mechanical properties.

* Try preparing a "flash card" showing the equations for the chemical reactions that take place in the blast furnace and another "flash card" showing the equations for the chemical reactions that take place in the basic oxygen furnace and memorizing them.

Copper Production

Copper-bearing ores include chalcopyrite ($CuFeS_2$), chalcocite (Cu_2S) and covellite (CuS). **The ores generally contain low percentages of copper and must be enriched by the process of flotation.** The ore is finely powdered, oil is added and the mixture is agitated with soapy water in a large tank (Fig. 1.18 in the text). Compressed air is forced through the mixture and the lightweight, oil-covered copper sulfide particles rise to the top as the heavier gangue particles settle on the bottom.

The enriched ore, typically chalcopyrite, is then **roasted** to convert the iron to iron(II) oxide and some of the sulfur to sulfur dioxide while leaving copper(II) sulfide. The mixture of iron and copper sulfides is then combined with sand and lime-stone and heated to 1100°C to decompose the limestone to lime (CaO) and carbon dioxide. The lime and iron(II) oxide then react with the silica (SiO_2) of sand to form calcium and iron silicate slag that is drawn off, and the excess sulfur reduces the copper(II) sulfide to copper(I) sulfide, Cu_2S, called *copper matte*.

The copper(I) sulfide is further heated in the presence of "blown air" and converted to copper metal and sulfur dioxide. **The copper that is obtained in this manner must be electrolytically purified for use in copper wiring. The impure copper is made the anode, pure copper is made the cathode and a solution of $CuSO_4$ and H_2SO_4 is used as the electrolyte. Metals more active than Cu, metals with more positive oxidation potentials, and copper are oxidized and enter into the solution at the anode while those less active fall to the bottom to form an "anode mud," that typically includes precious metals. The ions from the more active metals are harder to reduce than Cu^{2+}, and thus, remain in solution as Cu^{2+} ions are reduced to pure Cu metal at the cathode.**

* Try preparing a "flash card" showing the equations for the chemical reactions that take place in the pyrometallurgy of copper and memorizing it.

23.3 Coordination Compounds

When a metal salt dissolves in water, water molecules cluster around the ions. We commonly designate this by adding "aq" to the formula of these ions. The negative end of the polar water molecule is attracted to the positively charged metal ion (Fig. 17.12 in the text), and the positive end of the water molecule is attracted to the anion. Indeed, the energy of the ion-solvent interaction is what drives the solution process.

Hydrated ions and complex ions are examples of a large group of substances known as **coordination compounds,** which are species in which a metal atom or ion is associated with a group of neutral molecules or anions. *They may be neutral molecules, but species with an overall positive or negative charge are also included in this category, as are species containing main group metals instead of transition metals.*

Complexes and Ligands

The formula $FeCl_2 \cdot 6H_2O$ conveys the compound's stoichiometry but fails to give information about its structure. *The preferred method of writing the formula for a coordination compound places brackets around the metal atom or ion and the molecules or ions directly bonded to the metal.* Thus, it is preferable to write the formula for this compound as $[Fe(H_2O)_6]Cl_2$ to indicate that it is composed of a $[Fe(H_2O)_6]^{2+}$ cation and two chloride ions that are present to balance the charge in this ionic compound. The species $[Fe(H_2O)_6]^{2+}$ is called a **coordination complex**, or **complex ion.**

The molecules or ions attached to the metal ion are *called ligands, from the Latin verb ligare, meaning "to bind." Ligands have at least one atom that has a lone pair of electrons, and this lone pair gives the ligand the ability to bond to the metal. The classical view of the bonding in coordination complexes is that the lone pair of electrons on the ligand is shared with the metal in a Lewis acid-base interaction. The ligand is considered to be the Lewis base because it furnishes the electron pair and the metal is the Lewis acid* (Section 17.11 in the text.)

The number of ligand atoms attached to a metal is called the **coordination number**. As we shall see, *coordination complexes have a definite geometry or structure depending on the number of ligand atoms attached and the nature of the particular metal and ligands.* The situation is similar to our consideration of the VSEPR geometries of molecules (Section 9.5 in the text).

Ligands like H_2O and NH_3 have only one Lewis base atom that can attach to a metal and are classified as **monodentate**. Ligands that contain two or more atoms attached to a metal are classified as **polydentate**. Examples of polydentate ligands are given in Fig. 23.15 in the text. *Most polydentate ligands form five or six-membered rings with the metal allowing the atoms to maintain*

*their normal bond angles, because the 110° angles of a pentagon
are very close to the 109.5° tetrahedral bond angles and the 120°
angles of a hexagon agree with 120° triangular planar bond
angles.* ✶ Try preparing a "flash card" listing the names and formulas of the
bidentate ligands given in Fig. 23.15 in the text and a separate "flash card" for the
hexadentate EDTA^{4-} ion and memorizing them.

Polydentate ligands are also called **chelating ligands**, or
chelates (pronounced "key-late"). The name chelate is derived
from the Greek word *chele* meaning "claw." *Chelate complexes are
particularly stable because their formation involves a favorable
entropy change and more than one bond must be broken to separate
the ligand from the metal. The favorable entropy change for
forming a chelate complex is due to the increase in the number
of species and accompanying increase in disorder.* Consider,

$$[Ni(H_2O)_6]^{2+}(aq) + 3\ en(aq) \rightleftharpoons [Ni(en)_3]^{2+}(aq) + 6\ H_2O(\ell) \qquad (23a)$$

where en represents the bidentate ligand ethylenediamine, and
notice that 4 particles produce 7 particles and more disorder.

Exs. 23.1 and 23.2 in the text illustrate the use of the
formulas and charges of metals and ligands to write formulas of
coordination complexes and the use of formulas of coordination
complexes and the formulas and charges of ligands to determine
the oxidation numbers of the metals. *These are important skills
to be learned to be able to work with coordination compounds.*

Naming Coordination Compounds

*Coordination compounds are named following an internationally
established system.* The rules to be followed in naming coordination compounds are listed on pp. 1059 and 1060 in the text, and
their use is illustrated in Ex. 23.3 in the text. ✶ Try making a
photocopy of these rules and pasting them onto a "flash card" for use in naming
coordination compounds and writing the formulas of compounds given their names.

23.4 Structures of Coordination Compounds and Isomers

Common Geometries

*The geometry of a coordination complex is defined in terms of
the arrangement of the donor atoms of the ligands around the
central metal.* The metal can have any coordination number
between 2 and 12, but 2, 4, and 6 are most common, so we will
concentrate on species with these coordination numbers.

Complexes with the General Formula [ML$_2$]$^{n\pm}$

This stoichiometry is often encountered when the metal ion has +1 charge. Examples include CuCl$_2^-$ and Ag(NH$_3$)$_2^+$.

All complexes having this stoichiometry have a linear geometry. The two ligands are located on opposite sides of the metal, with a L-M-L bond angle of 180°.

Complexes with the General Formula [ML$_4$]$^{n\pm}$

Complexes having this stoichiometry would be expected to exist in a tetrahedral structure, as is observed for such complexes as [CoCl$_4$]$^{2-}$, [NiCl$_4$]$^{2-}$, [Zn(NH$_3$)$_4$]$^{2+}$ and Ni(CO)$_4$. *However, a large number of four-coordinate complexes are square-planar;* the ligands lie at the corners of a square that surrounds the metal and is in the same plane as the metal, so the L-M-L bond angles are 90°. *Square-planar complexes are often formed by metal ions that have the electron configuration [noble gas core](n-1)d^8.* This includes complexes of Pt^{2+} and Pd^{2+} and some of Ni^{2+}, such as [Ni(CN)$_4^{2-}$].

Complexes with the General Formula [ML$_6$]$^{n\pm}$

Complexes having this stoichiometry have the ligands arranged at the corners of an octahedron surrounding the metal, as illustrated in the margin figure on p. 1056 in the text.

Compounds having the same number of attached ligands from polydentate ligands also tend to have the geometries described above, as shown for [Co(en)$_3$]$^{3+}$ and [CoEDTA]$^-$ in Figs. 23.14 and 23.16 in the text.

Isomerism:

Molecules that have the same molecular formula but different bonding arrangements of atoms are called **structural isomers**, examples of which were encountered earlier with organic chemistry (11.1). In a second type of isomerism, **stereoisomerism**, the atom-to-atom bonding sequence is the same, but the atoms differ in their arrangement in space. *Two types of stereoisomerism occur.* One is **geometric isomerism**, in which the atoms making up the molecule are arranged in different geometrical relationships (11.1). *Cis*-2-butene and *trans*-2-butene shown on p. 504 of the text are examples of geometric isomers. The second type of stereoisomerism is **optical isomerism**, which arises when a molecule and its mirror image are not superimposable. Both geometric and optical isomerism are encountered in coordination chemistry.

Geometric Isomerism

Geometric isomers are obtained when the atoms bonded directly to the metal have a different spatial arrangement (different bond angles). The simplest example of geometric isomerism is cis-trans isomerism where cis means same side and trans means opposite side.

Cis-trans isomerism is not possible for tetrahedral complexes, because all ligand positions are adjacent to one another in a tetrahedron and all L-M-L bond angles are 109.5°. However, cis-trans isomerism can occur in both square planar and octahedral complexes. For example, in square-planar $Pt(NH_3)_2Cl_2$, the two Cl^- ligands can be adjacent to one another at 90° apart, be *cis* to one another, or be on opposite sides of the complex at 180° apart, be *trans* to one another, as shown on p.1062 in the text. Similarly, the two ammonia molecules in square-planar $Pt(NH_3)_2(Cl)(NO_2)$ can occupy *cis* or *trans* positions. Interestingly, *cis*-$Pt(NH_3)_2Cl_2$ is effective in the treatment of testicular, ovarian, bladder and osteogenic sarcoma cancers, but the *trans* isomer has no effect on these diseases (p. 7 in the text).

Cis-trans isomerism in octahedral complexes can be illustrated with $[Co(H_2NCH_2CH_2NH_2)_2Cl_2]^+$ which contains two bidentate ethylenediamine ligands and two Cl^- ligands and is octahedral. In this complex, the two Cl^- ligands can occupy either adjacent (*cis*) positions of opposite (*trans*) positions as shown in Fig. 23.18 in the text.

Another type of geometrical isomerism, called mer-fac isomerism, occurs with octahedral complexes having the general formula MX_3Y_3. In the fac isomer, three similar ligands lie at the corners of a triangular face of the octahedron (fac = facial), and in the mer isomer three similar ligands follow a meridian (mer = meridian), as illustrated for $Cr(NH_3)_3Cl_3$ in Fig. 23.19 in the text.

Optical Isomerism

Certain molecules have the same characteristic as hands and gloves: they exist in two mirror image structures that cannot be superimposed. Molecules that have nonsuperimposable mirror images are termed **chiral**, and molecules with superimposable mirror images are termed **achiral**. Nonsuperimposable molecules are known as **enantiomers.**

Enantiomers have the same stoichiometry and the same atom-to-atom bonding sequences, but they differ in the details of the arrangements of atoms in space. The most common example of chirality is found with carbon atoms bonded to four differing groups, as illustrated for lactic acid, $CH_3CH(OH)COOH$, in Fig. 23.21 in the text.

The two enantiomers have the same physical properties of melting point, boiling point, density and solubility in common solvents, but differ in their ability to rotate a beam of plane-polarized light. When a beam of plane-polarized light is passed through a solution of a pure enantiomer, the plane of polarization is twisted or rotated in one direction, as illustrated in Fig. 23.22 in the text. The two enantiomers rotate the plane-polarized light to the same extent but in opposite directions. Because of this, chiral compounds are referred to as optical isomers, and chiral compounds are said to be optically active.

Optical isomers of tetrahedral complexes are possible, but no commonly encountered examples of stable tetrahedral complexes are optically active. On the other hand, square-planar complexes are incapable of exhibiting optical activity based on the arrangements of the ligands around the metal, because all mirror images of this kind are superimposable.

There are numerous possibilities for chirality with octahedral complexes, but only one possibility is discussed in the text. In this case, a metal coordinates to three bidentate ligands, as in $[Co(en)_3]^{3+}$. The mirror images of this type of complex are shown in Fig. 23.23 in the text. Notice that because of the way that the five-membered rings are arranged, the mirror images cannot superimpose. Solutions of each optical isomer rotate polarized light in opposite directions whereas solutions containing equimolar concentrations of each isomer have no effect on polarized light.

23.5 Bonding in Coordination Compounds

Metal-ligand bonding was described earlier in this chapter in terms of the covalent bonding model of Lewis acid-base theory, the so-called valence bond theory model. However, this model is not adequate to explain the color and magnetism of coordination complexes and has been largely superseded by molecular orbital theory and crystal field theory.

Molecular orbital theory and crystal field theory both approach metal-ligand bonding using metal d orbitals and ligand lone pair orbitals. As the ligands approach to form bonds, two effects occur: (a) the metal and ligand orbitals overlap, and (b) the electrons of the metal repulse the electrons of the ligands. Molecular orbital theory takes both effects into account whereas the crystal field theory focuses on metal-ligand electron repulsions. the molecular orbital theory model assumes that the metal and ligands bond through molecular orbitals formed by atomic orbital overlaps between the metal and ligand. In contrast, the crystal field theory model assumes that the

positive charge of the metal and the negative charge of the
ligand lone pairs are attracted electrostatically. The two
theories, the molecular orbital theory and the crystal field
theory, produce the same qualitative results concerning the
color and magnetism of coordination complexes. The advantage of
the crystal field theory is that it is somewhat easier to
comprehend. The advantage of the molecular orbital theory is
that it is more complete and better able to give explanations
for certain phenomena, but it is also more difficult to comprehend, so we will focus on the crystal field theory approach.

d-Orbital Energies in Coordination Compounds

The shapes of the d orbitals, first shown in Fig. 7.17 in the
text, are repeated with in Fig. 23.24 in the text, where they
are arranged in two groups. The $d_{x^2-y^2}$ and d_{z^2} place their
electron densities along the x, y and z axes and are therefore
placed into one set whereas the d_{xy}, d_{xz} and d_{yz} orbitals place
their electron densities between the axes and are therefore
placed into a second set. We have chosen to divide the d
orbitals into these two sets, because we have chosen to assign
the ligands in square-planar and octahedral complexes to
positions along the x, y and z axes.

In an isolated atom or ion, the d orbitals are equal in energy
and are said to be **degenerate**. However, in an octahedral
complex, the d orbitals that are along the axes, the $d_{x^2-y^2}$ and
d_{z^2}, are pointed directly toward the electron pairs of the
ligands and are raised more in energy by metal d electron-ligand
electron pair repulsions than are the d orbitals that are
pointed between the axes, the d_{xy}, d_{xz} and d_{yz} orbitals. This
means an energy difference is created between the two sets of d
orbitals, as shown in Fig. 23.25 in the text. The energy
difference is designated Δ_o for octahedral, and it is a function
of the metal and the ligands and varies predictably from one
complex to another.

A different splitting pattern is obtained for square-planar
complexes, because the four ligands approach along only the x
and y axes. Thus, the d orbitals that are affected the most are
those having the greatest probabilities in the xy plane, with
the $d_{x^2-y^2}$ being affected most because it lies along the x and y
axes (Fig. 23.26 in the text). The next most affected d orbital
in a square-planar complex is the d_{z^2} which interacts at right
angles with all four ligands, and the least affected are the d_{xz}
and d_{yz} orbitals which each only interact indirectly with two
ligands.

Magnetic Properties of Coordination Compounds

The d orbital splitting in coordination complexes is the cause *of both the magnetic behavior and color for these species, and the keys to understanding and explaining these phenomenon are learning to correctly determine the number of d electrons for the metal and to correctly assign them to the split d orbitals.*

A gaseous or isolated Cr^{2+} ion has the electron configuration $[Ar]3d^4$. These electrons occupy four separate d orbitals in accord with Hund's rule giving four unpaired electrons. This occurs because the d orbitals for isolated atoms and ions are degenerate, but *in the presence of an octahedral ligand field, the d orbitals are split in energy, and there are two possible ways the electrons can be arranged. The first three electrons are necessarily assigned to the lower set of three d orbitals separately. However, the fourth electron can either be assigned to one of the two higher energy d orbitals or assigned to one of the three lower energy d orbitals where it would need to pair up with another electron.*

These two possibilities are shown as Configurations A and B, respectively in the margin figure on p. 1069 in the text and are also shown in Fig. 23.27 in the text. The first arrangement is called **high spin**, because it has the same number of unpaired electrons as the isolated metal atom or ion would have, the maximum number of unpaired electrons. The second arrangement is called **low spin** because it has less than the maximum number of unpaired electrons. *High-spin complexes arise whenever, the energy required for promoting an electron to one of the higher energy orbitals, Δ_o, is less than the energy that is required to overcome the repulsion between electrons occupying the same orbital and pair them,* the **pairing energy, P.** *Conversely, low-spin complexes arise whenever the promotion energy Δ_o exceeds the pairing energy P. That is,*

High-spin: $\Delta_o < P$

Low-spin: $\Delta_o > P$

The choice between high-spin and low-spin arises with octahedral complexes of metal atoms or ions having d^4 through d^7 electron configurations. The larger the value of Δ_o and the smaller the value of P, the more likely the complex will be low-spin.

The magnetic behavior of a coordination complex can be used to determine whether it is high-spin or low-spin, because of the difference in the number of unpaired electrons. In the case, of spare-planar complexes of Ni^{2+}, Pd^{2+} and Pt^{2+} having $(n-1)d^8$ electron configurations, it can be shown that only low-spin complexes are known. This indicates the four d orbitals lowest in energy are fully occupied by electrons having opposite spins.

23.6 The Colors of Coordination Compounds

The colors associated with coordination compounds are due to the selective absorption of the components of white, to the selective absorption of wavelengths within the visible portion of the electromagnetic spectrum (Fig. 7.3 in the text). *The color we observe is the complement of the colors absorbed*, as can be shown by using spectrophotometers like that outlined in Fig. 23.32 in the text.

The Absorption of Light by Coordination Complexes

The absorption of visible light is associated with the promotion of electrons to higher energy states. In the case of octahedral coordination compounds, this involves promotion from the lower set of d orbitals to the higher set of d orbitals. This is why the valence bond model is unable to account for the colors of coordination complexes, because it treats the d orbitals as being degenerate, as being equal in energy. *This is also why complexes of metals having d^0 or d^{10} electron configurations are colorless. In addition, this is why high spin complexes of metal ions having d^5 electron configurations are only faintly colored, because electron promotions involving changes of spin are not nearly as probable as those which do not involve changes of spin.*

The Spectrochemical Series of Ligands

The absorption spectra of complexes reveal that some ligands cause greater splittings of the d orbitals than others. A consideration of the spectra of some complexes of Co^{3+} (Table 23.4 in the text) and of other metals, suggests *the order of increasing ability to split the d orbitals is*

$$\text{Halides} < C_2O_4{}^{2-} < H_2O < NH_3 = en < phen < CN^- \qquad (23b)$$

small Δ_o large Δ_o
weak field ligands   strong field ligands

This order is called the **spectrochemical series** because it was determined by spectroscopy, and it is applicable to a wide range of metal complexes. *This means there is also an experimental basis for predicting which ligands are likely to cause metal complexes to be high-spin and which are likely to cause metal complexes to be low-spin. The so-called weak field ligands cause less splitting than the so-called strong field ligands and are therefore more likely to form high-spin complexes than are the strong field ligands. However, the pairing energy varies from metal-to-metal and is also a factor.*

23.7 Hints for Answering Questions and Solving Problems

Writing electron configurations for transition metal ions:

Be sure to remove ns electrons before $(n-1)d$ electrons from the [noble gas core]$ns^a(n-1)d^b$ configuration of the neutral atoms.

Predicting whether metals can be oxidized by 1 M H_3O^+:

Only metals having positive M/M^{2+} or M/M^{3+} E^O oxidation values can be oxidized by 1 M H_3O^+ to give the metal ions and hydrogen gas.

Predicting the geometries of $[ML_4]^{n\pm}$ complexes:

Four-coordinate complexes of metal ions having d^7 or d^{10} electron configurations are usually tetrahedral. In addition, many of the four coordinate complexes of d^8 Ni^{2+} are tetrahedral owing to its small size and the opportunity for 109.5° bond angles compared to 90° bond angles for square-planar complexes. (The only common exception is $Ni(CN)_4^{2-}$.) On the other hand, four-coordinate complexes of larger ions having d^8 configurations, such as Pd^{2+}, Ir^+, Pt^{2+} and Au^{3+}, are commonly square-planar complexes.

Predicting whether octahedral complexes are likely to be low-spin or high-spin complexes:

The only ligand listed in the spectrochemical series given in the text that consistently forms low-spin complexes is CN^-. The other ligands are most likely to form high-spin complexes with the M^{2+} and M^{3+} ions of elements of the first transition series, except Co^{3+}. Co^{3+} forms high-spin complexes with the halide ions and low-spin complexes with all the other ligands of the spectrochemical series.

Predicting whether coordination complexes are diamagnetic or paramagnetic:

Paramagnetic substances have unpaired electrons (p. 358 in the text), so the keys to correctly predicting whether a coordination complex is diamagnetic or paramagnetic are 1) correctly determining the charge of the metal ion and the number of d electrons for the metal ion and 2) correctly placing the appropriate number of d electrons into the two sets of d orbitals according to whether the complex is known or predicted to be high-spin or low-spin.

CHAPTER 24

NUCLEAR CHEMISTRY

The area of nuclear chemistry is a fascinating area because it deals with the study of a phenomenon that has an amazing range of applications, some of which are knowingly harmful to some of which are extremely useful and valuable. For example, no one can deny the harm of nuclear weapons or the value of the energy from the sun that is due to nuclear reactions. Indeed, no topic could be more appropriate and relevant for the conclusion of your study of general chemistry than nuclear chemistry.

24.1 The Nature of Radioactivity

By the early part of this century, it was known that there were at least three basic kinds of radiation:

- **Alpha (α) radiation** consists of particles that are attracted to electrically negative plates and are now known to be $_2^4He^{2+}$ ions (i.e. helium nuclei). Alpha particles are quite massive and can be stopped by several sheets of ordinary paper or clothing.

- **Beta (β) radiation** consists of particles that are attracted to electrically positive plates and are now known to be electrons. They are much lighter and more penetrating than alpha particles. Beta particles can penetrate several millimeters of living bone or tissue, and least ⅛ inch of aluminum is required to stop beta particles.

- **Gamma (γ) radiation** is a form of electromagnetic radiation that has no electrical charge and is not affected by an electrical field. Gamma radiation can pass completely through the human body, and thick layers of lead or concrete are required to stop gamma radiation.

The characteristics of α, β and γ emissions are summarized in Table 24.1 in the text. ✴ Try preparing a "flash card" summarizing the information of Table 24.1 and memorizing it.

24.2 Nuclear Reactions

In 1902, Ernest Rutherford and Frederick Soddy suggested *radioactivity is the result of a natural change of the isotope of one element to the isotope of a different element.* In these changes, called **nuclear reactions** or **transmutations**, an unstable nucleus emits radiation and is converted into a more stable nucleus of a different element. *Thus, there is a change in atomic number and often a change in mass number, as well.*

The total number of nuclear particles, **nucleons** (protons plus neutrons), remains the same during nuclear reactions. *A proton can change to a neutron or a neutron can change to a proton*, but *the sum of the mass numbers of the reacting nuclei must equal the sum of the mass numbers of the nuclei produced. Furthermore, to maintain charge balance, the sum of the atomic numbers of the products must equal the sum of the atomic numbers of the reactants.*

Reactions Involving Alpha and Beta Particles

In alpha emission the atomic number of the emitting atom decreases by two units and the mass number decreases by four units for each alpha particle emitted. Thus,

$$^{214}_{84}Po \rightarrow {}^{210}_{82}Pb + {}^{4}_{2}He \tag{24a}$$

In beta emission the atomic number of the emitting atom increases by one for each beta particle emitted, but the mass number does not change. Thus,

$$^{14}_{6}C \rightarrow {}^{14}_{7}N + {}^{0}_{-1}\beta \tag{24b}$$

There are no electrons in the nucleus, but beta particle emission is thought to occur as a neutron is converted to a proton in a series of steps. The net equation for this change is

$$^{1}_{0}n \rightarrow {}^{1}_{1}p + {}^{0}_{-1}\beta \tag{24c}$$

In some cases, the isotopes formed by alpha and beta emission are radioactive, and a number of successive transformations occur to give a stable, nonradioactive isotope in a **radioactive series.** One such series starts with uranium-238 and ends with lead-206 (Fig. 24.2 in the text).

Other Types of Radioactive Decay

Some nuclei decay by emitting a **positron**, $_{-1}^{0}\beta$, which is effectively a positively charged electron. *Positron emission is therefore the opposite of beta emission, and the atomic number of the emitting atom decreases by one for each positron emitted and the mass number does not change.* Thus,

$$_{11}^{20}Na \rightarrow _{10}^{20}Ne + _{+1}^{0}\beta \tag{24d}$$

Positrons are sometimes called "antielectrons", because collisions with electrons lead to the mutual annihilation of both particles with the production of two high energy gamma rays.

Other nuclei decay by **electron capture**. *In this process an inner shell electron is captured by the nucleus, and a proton is converted to a neutron.*

$$_{1}^{1}p + _{-1}^{0}e(\text{orbital electron}) \rightarrow _{0}^{1}n \tag{24e}$$

The atomic number of the emitting atom therefore decreases by one for each electron captured, but the mass number does not change. Thus,

$$_{23}^{50}V + _{-1}^{0}e(\text{orbital electron}) \rightarrow _{22}^{50}Ti \tag{24f}$$

Electron capture is sometimes called **K-capture**, because the innermost shell was called the K-shell in the old nomenclature of atomic physics.

Summary:

The net effects of the four most common modes of radioactive decay are summarized in the unnumbered figure on p. 1092 in the text. Notice that it is common practice to call the isotope that begins the radioactive decay process the **parent** and the isotope that is produced the **daughter**. ✶ Try preparing a "flash card" summarizing the changes in atomic number and mass number that occur during alpha, beta and positron emissions and memorizing it.

24.3 Stability of Atomic Nuclei

Hydrogen exists in three isotopic forms. In its simplest and most abundant form called **protium**, $_{1}^{1}H$, hydrogen has one proton as its only nuclear particle. In **deuterium**, $_{1}^{2}H$ or $_{1}^{2}D$, hydrogen has one proton and one neutron as nuclear particles, and in radioactive **tritium**, $_{1}^{3}H$ or $_{1}^{3}T$, hydrogen has one proton and two neutrons as nuclear particles.

Most elements exist in more than one isotopic form, and with the exception of $_1^1H$ and $_2^3He$, all these isotopes have mass numbers that are at least twice as large as their atomic number. This means there is at least one neutron for every proton and implies that neutrons are able to reduce the repulsive forces between the protons. Consider the following characteristics of stable nuclei gleaned from Fig. 24.3 in the text showing stable nuclei as a function of number of neutrons vs. number of protons:

- For elements through $_{20}Ca$, stable isotopes usually have equal numbers of neutrons and protons (i.e. N = Z), or perhaps one extra neutron.

- For elements beyond $_{20}Ca$, the band of stable isotopes deviates more and more from the N = Z line indicating more and more neutrons are needed per proton for nuclear stability.

- Beyond $_{83}Bi$ all isotopes are unstable. Furthermore, the rate of disintegrations becomes greater the heavier the nuclei.

- Elements of even atomic numbers have more stable isotopes than do those with odd atomic numbers. Secondly, stable isotopes generally have an even number of neutrons. Thus, stable nuclei are likely to have even numbers of both protons and neutrons. Indeed, of the more than 300 isotopes shown in the band of stability shown in Fig. 24.3 in the text

  - approximately 200 have even numbers of both protons and neutrons,

  - approximately 120 have an even number of either protons or neutrons, and

  - only four ($_1^2H$, $_3^6Li$, $_5^{10}B$ and $_7^{14}N$) have odd numbers of both protons and neutrons.

The Band of Stability and Type of Radioactive Decay

Any isotope not on the "band of stability" shown in Fig. 24.3 in the text decays in such a way that it can come closer to it.

  - Beta emission occurs when isotopes have too many neutrons and are above the "band of stability."

- Positron emission or electron capture occurs when isotopes have too many protons and are below the "band of stability."

- Alpha emission often occurs with elements beyond $_{83}Bi$.

✳ Try preparing a "flash card" showing the general form of Fig. 24.3 and label the three regions just described and memorizing it.

Binding energy

A measure of the short-range forces holding nuclei together is given by the nuclear binding energy. The **binding energy, E_b,** of a nucleus is defined as the negative of the energy change (ΔE) that would occur if a nucleus were formed from its constituent protons and neutrons. **Nuclear binding energies are typically huge positive quantities ($\approx 10^8$ kJ/mol nucleons) indicating the forces holding nuclei together are indeed very strong.**

What is the source of this nuclear binding energy? It can be shown experimentally that the masses of atoms other than $_1^1H$ are always less than the sum of the masses of its constituent protons and neutrons. The theory is that the "missing mass," Δm, sometimes called the **mass defect**, is converted to energy that binds the nucleus together.

The relation between mass and energy is found in Albert Einstein's 1905 theory of special relativity. Einstein contends mass and energy are simply different manifestations of the same quantity and can therefore be converted into one another. Einstein stated that the energy of a body is equivalent to its mass times the square of the speed of light, $E = mc^2$. Thus, the energy change that occurs when mass is converted to energy is given by

$$\Delta E = (\Delta m) c^2 \tag{24g}$$

where ΔE is calculated in joules when the mass is expressed in kilograms and the velocity of light is expressed in meters per second, because $1 \text{ J} = 1 \text{ kg} \cdot \text{m}^2/\text{s}^2$. To be able to compare the stability of one nucleus to another, scientists calculate the **binding energy per nucleon,** binding energy per nuclear particle.

The greater the binding energy per nucleon, the greater the stability of the nucleus. Thus, $_{26}^{56}Fe$ appears at the maximum in the plot of binding energies vs. atomic number given in Fig. 24.4 in the text and is considered to have the most stable nucleus. *This is the reason that iron is the most abundant of the heavy elements in the universe.*

24.4   Rates of Disintegration Reactions

Radioactive decay processes are first-order processes like those encountered in Chapter 15. The rates at which radioactive isotopes used in medicine decay is of interest to us, as are the rates of decay of radioactive pollutants. Furthermore, decay rates are used to determine the age of objects, including the earth.

Half-Life

The time required for one-half of a radioactive isotope to decay is called its **half-life, $t_{1/2}$.** The half-lives of radioactive isotopes range from fractions of seconds to more than a million years and are taken to be measures of the relative stabilities of such isotopes. This can be done because half-lives for radioactive decay reactions are independent of temperature and the number of radioactive nuclei present. **The fraction of an initial amount of a radioactive isotope that is present after n half-lives is always equal to $(1/2)^n$.**

Rate of Radioactive Decay

The rate of decay must be measured to determine the half-life of a radioactive isotope. The rate of decay is usually described in terms of the **activity (A)** of the sample, the number of disintegrations occurring per unit time. *The activity is proportional to the number of radioactive atoms present (N), so*

$$A = kN \qquad (24h, 24.1)$$

where k is the first order rate constant called the **decay constant**.

The equation that relates the number of radioactive nuclei at any point in the decay to the original number of radioactive nuclei, time of decay and decay constant is the integrated rate law for first order processes expressed in terms of number of radioactive nuclei:

$$\ln(N/N_0) = -kt \qquad (24i, 24.2)$$

where N/N_0 is the fraction of radioactive atoms remaining from an original sample after time t, the number of radioactive atoms

N that are present at time t divided by the number present at the beginning of the experiment N_0.

Because activity (A) is proportional to the number of radioactive nuclei (N), Expression 24b can also be written as

$$\ln(A/A_0) = -kt \qquad (24j)$$

The advantage of this form is that it is expressed in terms of experimental information, because radioactive activities can be measured using devices such as Geiger counters (Fig. 24.6 in the text).

The half-life expression for radioactive decay is the usual half-life expression for first-order processes (Chapter 15):

$$t_{1/2} = \frac{0.693}{k} \qquad (24k, 24.3)$$

Thus, **Expressions 24b and c can be used in three ways**:

- The ratio A/A_0 can be determined by measuring the activity of a sample over some time period, t, and the value of k can be calculated. The value of k can then be used with Expression 24d to calculate the value of the half-life of the isotope, as illustrated in Ex. 19.3 in the text.

- If the value of k is known from measuring the activity of a sample over some time period, t, it can be used to calculate the fraction of any radioactive sample of the same isotope that will remain (N/N_0) after some time t has elapsed.

- If the value of k is known from measuring the activity of a sample over some time period, t, it can be used to calculate the time that will be required for that isotope to decay from an initial activity (A_0) to a certain activity (A), as illustrated in Exs. 19.4 and 19.5 in the text.

Radiochemical Dating

In 1929, physicist Serge Korff discovered that carbon-14 is continually generated in the upper atmosphere. High-energy cosmic rays smash into gases causing them to eject neutrons, which collide with nitrogen atoms to produce carbon-14.

$$^1_0n + {}^{14}_7N \rightarrow {}^{14}_6C + {}^1_1H \qquad (24\ell)$$

Chapter 24

It is estimated that only 7.5 kg of ^{14}C is produced annually by this reaction, but it gets mixed throughout the atmosphere and becomes part of the carbon intake of all living organisms.

All living organisms contain ^{12}C, ^{13}C and ^{14}C, but only ^{14}C is radioactive. It decays by beta emission

$$^{14}C \rightarrow {}^{14}N + {}_{-1}^{0}\beta \qquad (24m)$$

and has a half-life of 5.73 X 10^3 years.

In 1946, Willard F. Libby realized that the intake of carbon-14 ceases when an organism dies and that the carbon-14 activity of objects derived from living organisms could be used to determine the dates of these objects using Expression 24c. The assumption is made that the carbon-14 content of the atmosphere has remained constant and that the original carbon-14 activity of the object is therefore equal to the carbon-14 activity of current samples of the organism. For example, the carbon-14 activity of a newly prepared piece of paper could be used to determine the age of a paper object or the carbon-14 activity of a living tree could be used to determine the age of a wooden object, assuming the object was made as soon as the lumber was prepared. Ex. 24.5 in the text contains an example based on radiocarbon dating.

The technique of carbon-14 dating is limited to objects that are less than 30,000-50,000 years old, because the ^{14}C activity becomes too low to be measured accurately. The dates obtained by carbon-14 dating calculations must also be corrected by correlation with results of other techniques, such tree-ring counting. This is due to changes in the solar activities that produce ^{14}C, changes in the earth's magnetic field that influences the earth's ability to capture cosmic rays and the huge influx of carbon atoms into the atmosphere through the burning of fossil fuels, which are naturally depleted of ^{14}C as a result of decay with time. Nevertheless, carbon-14 dating remains a useful dating technique.

24.5 Artificial Transmutations

In 1919, Ernest Rutherford discovered that bombardment of nitrogen atoms with alpha particles produced oxygen atoms

$$^4_2He + {}^{14}_7N \rightarrow {}^{17}_8O + {}^1_1H \qquad (24n)$$

and in 1932, James Chadwick discovered the neutron as a result of bombardment of beryllium atoms with alpha particles.

$$\text{}_2^4\text{He} + \text{}_4^9\text{Be} \rightarrow \text{}_6^{12}\text{C} + \text{}_0^1\text{n} \qquad (24o)$$

Reactions which involve bombarding stable nuclei with particles to cause nuclear reactions to occur to produce other elements are called **artificial transmutation reactions**.

Enrico Fermi reasoned that it would be easier to bombard elements with electrically neutral neutrons than alpha particles, because of the repulsive forces between the positively charged alpha particles and the protons in the nucleus of the targeted atom. Thus, neutron bombardment has been used to prepare elements beyond uranium in the periodic table, the so-called transuranium elements. However, elements beyond 101 have been prepared using heavier particles and specially designed particle accelerators for bombardment. The most recent of these discoveries/preparations (1994) are described the "Current Issues in Chemistry" article on p. 1105 in the text.

24.6 Nuclear Fission

In 1938, Otto Hahn and Fritz Strassman discovered some barium in a sample of uranium that had been bombarded with neutrons. Further work by Lise Meitner, Otto Frisch, Neils Bohr and Leo Sziland confirmed that the ^{235}U had captured a neutron to form ^{236}U, and ^{236}U had split in two in a **nuclear fission reaction**:

$$^{235}\text{U} + \text{}_0^1\text{n} \rightarrow \text{}^{236}\text{U} \rightarrow \text{}_{56}^{141}\text{Ba} + \text{}_{36}^{92}\text{Kr} + 3\text{}_0^1\text{n} \quad \Delta E = -2 \times 10^{10} \text{ kJ/mol} \quad (24p)$$

The fact that a fission produces more neutrons than it consumes is important, because if there is a critical mass of ^{235}U present to absorb the excess neutrons, the reaction can take place in an explosive **chain reaction** (Fig. 24.9 in the text).

In an atomic bomb, two small pieces of less than critical mass are brought together to form a single piece capable of sustaining a chain reaction that results in an explosion. In a **nuclear reactor** like those used electric power plants, the percentage of fissionable ^{235}U is much smaller than in an atomic bomb ($\approx$3% vs. $\approx$100%) and the rate of fission is controlled by withdrawing or inserting cadmium rods or other neutron absorbers into the reactor. Thus, there is no chance of having an explosion, but there is the chance of having a meltdown, and there is also the problem of disposing of the radioactive fission products.

24.7 Nuclear Fusion

The energy that is released when light nuclei combine to form a heavier nucleus in a **nuclear fusion** reaction can exceed the energy produced by a fission reaction. *One of the best known fusion reactions occurs in the sun where hydrogen nuclei fuse together to form helium nuclei:*

$$4 \; {}^1_1H \rightarrow {}^4_2He + 2 \; {}^0_{+1}\beta \qquad \Delta E = -2.5 \times 10^9 \text{ kJ} \qquad (24q)$$

Temperatures of 10^6 to 10^7 K, found in the core and radiative zone of the sun, are required to bring the positively charged nuclei together with enough kinetic energy to overcome the nuclear repulsions. Attempts to conduct fusion reactions for peaceful uses have thus far been unsuccessful, though fusion was the source of energy with the hydrogen bomb.

24.8 Radiation Effects and Units of Radiation

The following units are commonly used to quantify radiation and its effects on humans:

- The **röntgen (R)** is used to give the dosage of x-rays and γ rays. One röntgen corresponds to the deposition of 93.3 $\times 10^{-7}$ J per gram of tissue.

- The **rad** measures the amount of radiation absorbed. One rad corresponds to 1.00×10^{-5} J absorbed per gram of material.

- The **rem** (standing for röntgen equivalent man) is used to quantify the biological effects of radiation. One rem corresponds to any dose of radiation that has the effect of 1 R.

- The **curie** (Ci) is used as a unit of activity. One curie corresponds to 3.7×10^{10} disintegrations per second.

Humans are constantly exposed to natural and artificial **background radiation,** as noted in Table 24.3 in the text.

Radon

Radon is a chemically inert gas that occurs naturally in our environment. When inhaled by humans, ${}^{222}Rn$ can decay inside the lungs to give chemically active and radioactive ${}^{218}Po$ through alpha emission and ${}^{218}Po$ can further decay by alpha emission to

---

give radioactive ^{214}Pb. The alpha particles can damage the cells of the lungs and induce lung cancer, so it is wise to have the level of radon tested in homes, especially in areas known to have relatively high levels of radon.

24.9 Applications of Radioactivity

Food Irradiation

Foods can be pasteurized by irradiation with gamma rays from sources such as ^{60}Co and ^{137}Cs to retard the growth of organisms such as bacteria, molds and yeast. This sterilizes the organisms and prolongs the shelf-life of the foods, such as those listed in Table 24.4 in the text.

Radioactive Tracers

Chemists can use radioisotopes as **tracers** in chemical and biological processes. Compounds are prepared using radioactive atoms in place of stable atoms, and the course of the radioactive atoms can be determined by using a Geiger counter or similar instrument. Radioisotopes can also be used as tracers to locate leaks in underground pipes used to transport liquids or gases.

Medical Imaging

Radioactive isotopes are also used in **nuclear medicine** for diagnosis and therapy. *Certain isotopes selectively accumulate in damaged tissues and can be used to locate these areas*, as noted by the information given in Table 24.6 in the text. The nuclear chemistry of **positron emission tomography** (PET Scan Imaging) is also described in the text.

24.10 Key Expressions

Einstein's mass - energy expression:

$$E = mc^2$$

Nuclear binding energy - mass defect expression:

$$\Delta E = (\Delta m) c^2$$

First-order nuclear decay expressions:

$$\ln(N/N_o) = -kt$$

$$\ln(A/A_o) = -kt$$

First-order half-life expression:

$$t_{1/2} = \frac{0.693}{k}$$

24.11 Hints for Answering Questions and Solving Problems

Predicting the type of spontaneous radioactive decay:

The mode of radioactive decay is easiest to predict when the atomic number of the isotope is less than 20 or greater than 83. If it is less than 20, the isotope should have 1 proton: 1 neutron, or at most 1 extra neutron. If it has too many neutrons, it lies above the "belt of stability" and is a beta particle emitter. If it has too many protons, it lies below the "belt of stability" and is most likely a positron emitter, though it could undergo electron capture. If the isotope has an atomic number >83 and a mass number >200, it is most likely an alpha particle emitter, though it could be a beta particle emitter.

Writing balanced equations for nuclear reactions:

Make sure the sum of the mass numbers on the left side of the equation equals the sum of the mass numbers on the right side, and also make sure the sum of the atomic numbers on the left side equals the sum of the atomic numbers on the right side. If there is more than one kind of particle on either side, make sure you multiply the atomic number and mass number of the particle by its coefficient. This means 3 $_0^1n$ gives a total of 3 toward the sum of the mass numbers on its side of the equation.

Discerning between fission and fusion:

Fission involves the splitting of one heavy nucleus into two lighter nuclei whereas fusion involves the joining of two light nuclei to give one heavier nucleus.

INDEX